Markov Logic:
An Interface Layer for
Artificial Intelligence

Synthesis Lectures on Artificial Intelligence and Machine Learning

Editors
Ronald J. Brachman, *Yahoo! Research*
Thomas Dietterich, *Oregon State University*

Markov Logic: An Interface Layer for Artificial Intelligence
Pedro Domingos and Daniel Lowd
2009

Introduction to Semi-Supervised Learning
Xiaojin Zhu and Andrew B. Goldberg
2009

Action Programming Languages
Michael Thielscher
2008

Representation Discovery using Harmonic Analysis
Sridhar Mahadevan
2008

Essentials of Game Theory: A Concise Multidisciplinary Introduction
Kevin Leyton-Brown, Yoav Shoham
2008

A Concise Introduction to Multiagent Systems and Distributed Artificial Intelligence
Nikos Vlassis
2007

Intelligent Autonomous Robotics: A Robot Soccer Case Study
Peter Stone
2007

Markov Logic: An Interface Layer for Artificial Intelligence
Pedro Domingos and Daniel Lowd

ISBN: 978-3-031-00421-6   paperback
ISBN: 978-3-031-01549-6   ebook

DOI 10.1007/978-3-031-01549-6

A Publication in the Springer series
SYNTHESIS LECTURES ON ARTIFICIAL INTELLIGENCE AND MACHINE LEARNING

Lecture #7
Series Editors: Ronald J. Brachman, *Yahoo! Research*
             Thomas Dietterich, *Oregon State University*

Series ISSN
Synthesis Lectures on Artificial Intelligence and Machine Learning
Print 1939-4608   Electronic 1939-4616

Markov Logic:
An Interface Layer for Artificial Intelligence

Pedro Domingos and Daniel Lowd

University of Washington, Seattle

With contributions from Jesse Davis, Tuyen Huynh, Stanley Kok, Lilyana Mihalkova, Raymond J. Mooney, Aniruddh Nath, Hoifung Poon, Matthew Richardson, Parag Singla, Marc Sumner, and Jue Wang

SYNTHESIS LECTURES ON ARTIFICIAL INTELLIGENCE AND MACHINE LEARNING #7

ABSTRACT

Most subfields of computer science have an interface layer via which applications communicate with the infrastructure, and this is key to their success (e.g., the Internet in networking, the relational model in databases, etc.). So far this interface layer has been missing in AI. First-order logic and probabilistic graphical models each have some of the necessary features, but a viable interface layer requires combining both. Markov logic is a powerful new language that accomplishes this by attaching weights to first-order formulas and treating them as templates for features of Markov random fields. Most statistical models in wide use are special cases of Markov logic, and first-order logic is its infinite-weight limit. Inference algorithms for Markov logic combine ideas from satisfiability, Markov chain Monte Carlo, belief propagation, and resolution. Learning algorithms make use of conditional likelihood, convex optimization, and inductive logic programming. Markov logic has been successfully applied to problems in information extraction and integration, natural language processing, robot mapping, social networks, computational biology, and others, and is the basis of the open-source Alchemy system.

KEYWORDS

Markov logic, statistical relational learning, machine learning, graphical models, first-order logic, probabilistic logic, Markov networks, Markov random fields, inductive logic programming, satisfiability, Markov chain Monte Carlo, belief propagation, collective classification, link prediction, link-based clustering, entity resolution, information extraction, social network analysis, natural language processing, robot mapping, computational biology

Contents

Acknowledgments

We are grateful to all the people who contributed to the development of Markov logic and Alchemy: colleagues, users, developers, reviewers, and others. We thank our families for their patience and support.

The research described in this book was partly funded by ARO grant W911NF-08-1-0242, DARPA contracts FA8750-05-2-0283, FA8750-07-D-0185, HR0011-06-C-0025, HR0011-07-C-0060 and NBCH-D030010, NSF grants IIS-0534881, IIS-0803481 and EIA-0303609, ONR grants N-00014-05-1-0313 and N00014-08-1-0670, an NSF CAREER Award (first author), a Sloan Research Fellowship (first author), an NSF Graduate Fellowship (second author) and a Microsoft Research Graduate Fellowship (second author). The views and conclusions contained in this document are those of the authors and should not be interpreted as necessarily representing the official policies, either expressed or implied, of ARO, DARPA, NSF, ONR, or the United States Government.

Pedro Domingos and Daniel Lowd
Seattle, Washington

CHAPTER   1

Introduction

1.1   THE INTERFACE LAYER

Artificial intelligence (AI) has made tremendous progress in its first 50 years. However, it is still very far from reaching and surpassing human intelligence. At the current rate of progress, the crossover point will not be reached for hundreds of years. We need innovations that will permanently increase the rate of progress. Most research is inevitably incremental, but the size of those increments depends on the paradigms and tools that researchers have available. Improving these provides more than a one-time gain; it enables researchers to consistently produce larger increments of progress at a higher rate. If we do it enough times, we may be able to shorten those hundreds of years to decades, as illustrated in Figure 1.1.

If we look at other subfields of computer science, we see that in most cases progress has been enabled above all by the creation of an interface layer that separates innovation above and below it, while allowing each to benefit from the other. Below the layer, research improves the foundations (or, more pragmatically, the infrastructure); above it, research improves existing applications and invents new ones. Table 1.1 shows examples of interface layers from various subfields of computer science. In each of these fields, the development of the interface layer triggered a period of rapid progress above and below it. In most cases, this progress continues today. For example, new applications enabled by the Internet continue to appear, and protocol extensions continue to be proposed. In many cases, the progress sparked by the new layer resulted in new industries, or in sizable expansions of existing ones.

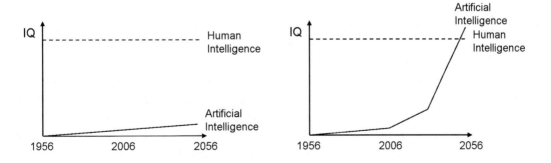

Figure 1.1: The first 100 years of AI. The graph on the right illustrates how increasing our rate of progress several times could bring about human-level intelligence much faster.

Table 1.1: Examples of interface layers.

Field	Interface Layer	Below the Layer	Above the Layer
Hardware	VLSI design	VLSI modules	Computer-aided chip design
Architecture	Microprocessors	ALUs, buses	Compilers, operating systems
Operating systems	Virtual machines	Hardware	Software
Programming systems	High-level languages	Compilers, code optimizers	Programming
Databases	Relational model	Query optimization, transaction mgmt.	Enterprise applications
Networking	Internet	Protocols, routers	Web, email
HCI	Graphical user interface	Widget toolkits	Productivity suites
AI	???	Inference, learning	Planning, NLP, vision, robotics

The interface layer allows each innovation below it to automatically become available to all the applications above it, without the "infrastructure" researchers having to know the details of the applications, or even what all the existing or possible applications are. Conversely, new applications (and improvements to existing ones) can be developed with little or no knowledge of the infrastructure below the layer. Without it, each innovation needs to be separately combined with each other, an $O(n^2)$ problem that in practice is too costly to solve, leaving only a few of the connections made. When the layer is available, each innovation needs to be combined only with the layer itself. Thus, we obtain $O(n^2)$ benefits with $O(n)$ work, as illustrated in Figure 1.2.

In each case, the separation between applications above the layer and infrastructure below it is not perfect. If we care enough about a particular application, we can usually improve it by developing infrastructure specifically for it. However, most applications are served well enough by the general-purpose machinery, and in many cases would not be economically feasible without it. For example, ASICs (application-specific integrated circuits) are far more efficient for their specific applications than general-purpose microprocessors; but the vast majority of applications are still run on the latter. Interface layers are an instance of the 80/20 rule: they allow us to obtain 80% of the benefits for 20% of the cost.

The essential feature of an interface layer is that it provides a language of operations that is all the infrastructure needs to support, and all that the applications need to know about. Designing it is a difficult balancing act between providing what the applications need and staying within what the infrastructure can do. A good interface layer exposes important distinctions and hides unimportant ones. How to do this is often far from obvious. Creating a successful new interface layer is thus not easy. Typically, it initially faces skepticism, because it is less efficient than the existing alternatives, and appears too ambitious, being beset by difficulties that are only resolved by later research. But once the layer becomes established, it enables innovations that were previously unthinkable.

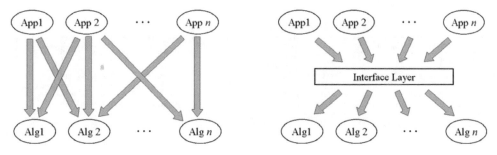

Figure 1.2: An interface layer provides the benefits of $O(n^2)$ connections between applications and infrastructure algorithms with just $O(n)$ connections.

1.2   WHAT IS THE INTERFACE LAYER FOR AI?

In AI, the interface layer has been conspicuously absent, and this, perhaps more than any other factor, has limited the rate of progress. An early candidate for this role was first-order logic. However, it quickly became apparent that first-order logic has many shortcomings as a basis for AI, leading to a long series of efforts to extend it. Unfortunately, none of these extensions has achieved wide acceptance. Perhaps the closest first-order logic has come to providing an interface layer is the Prolog language. However, it remains a niche language even within AI. This is largely because, being essentially a subset of first-order logic, it shares its limitations, and is thus insufficient to support most applications at the 80/20 level.

Many (if not most) of the shortcomings of logic can be overcome by the use of probability. Here, graphical models (i.e., Bayesian and Markov networks) have to some extent played the part of an interface layer, but one with a limited range. Although they provide a unifying language for many different probabilistic models, graphical models can only represent distributions over propositional universes, and are thus insufficiently expressive for general AI. In practice, this limitation is often circumvented by manually transforming the rich relational domain of interest into a simplified propositional one, but the cost, brittleness, and lack of reusability of this manual work is precisely what a good interface layer should avoid. Also, to the extent that graphical models can provide an interface layer, they have done so mostly at the conceptual level. No widely accepted languages or standards for representing and using graphical models exist today. Many toolkits with specialized functionality have been developed, but none that could be used as widely as (say) SQL engines are in databases. Perhaps the most widely used such toolkit is BUGS, but it is quite limited in the learning and inference infrastructure it provides. Although very popular with Bayesian statisticians, it fails the 80/20 test for AI.

It is clear that the AI interface layer needs to integrate first-order logic and graphical models. One or the other by itself cannot provide the minimum functionality needed to support the full range of AI applications. Further, the two need to be fully integrated, and not simply provided alongside each other. Most applications require simultaneously the expressiveness of first-order logic and the

Table 1.2: Examples of logical and statistical AI.

Field	Logical Approach	Statistical Approach
Knowledge representation	First-order logic	Graphical models
Automated reasoning	Satisfiability testing	Markov chain Monte Carlo
Machine learning	Inductive logic programming	Neural networks
Planning	Classical planning	Markov decision processes
Natural language processing	Definite clause grammars	Probabilistic context-free grammars

robustness of probability, not just one or the other. Unfortunately, the split between logical and statistical AI runs very deep. It dates to the earliest days of the field, and continues to be highly visible today. It takes a different form in each subfield of AI, but it is omnipresent. Table 1.2 shows examples of this. In each case, both the logical and the statistical approach contribute something important. This justifies the abundant research on each of them, but also implies that ultimately a combination of the two is required.

In recent years, we have begun to see increasingly frequent attempts to achieve such a combination in each subfield. In knowledge representation, knowledge-based model construction combines logic programming and Bayesian networks, and substantial theoretical work on combining logic and probability has appeared. In automated reasoning, researchers have identified common schemas in satisfiability testing, constraint processing and probabilistic inference. In machine learning, statistical relational learning combines inductive logic programming and statistical learning. In planning, relational MDPs add aspects of classical planning to MDPs. In natural language processing, work on recognizing textual entailment and on learning to map sentences to logical form combines logical and statistical components. However, for the most part these developments have been pursued independently, and have not benefited from each other. This is largely attributable to the $O(n^2)$ problem: researchers in one subfield can connect their work to perhaps one or two others, but connecting it to all of them is not practically feasible.

1.3   MARKOV LOGIC AND ALCHEMY: AN EMERGING SOLUTION

We have recently introduced Markov logic, a language that combines first-order logic and Markov networks. A knowledge base (KB) in Markov logic is a set of first-order formulas with weights. Given a set of constants representing objects in the domain of interest, it defines a probability distribution over possible worlds, each world being an assignment of truth values to all possible ground atoms. The distribution is in the form of a log-linear model: a normalized exponentiated weighted combination of features of the world.[1] Each feature is a grounding of a formula in the KB, with the corresponding weight. In first-order logic, formulas are hard constraints: a world that violates even a single formula

---

[1]Log-linear models are also known as, or closely related to, Markov networks, Markov random fields, maximum entropy models, Gibbs distributions, and exponential models; and they have Bayesian networks, Boltzmann machines, conditional random fields, and logistic regression as special cases.

Table 1.3: A comparison of Alchemy, Prolog, and BUGS.

Aspect	Alchemy	Prolog	BUGS
Representation	First-order logic + Markov networks	Horn clauses	Bayesian networks
Inference	Model checking, MCMC, lifted BP	Theorem proving	Gibbs sampling
Learning	Parameters and structure	No	Parameters
Uncertainty	Yes	No	Yes
Relational	Yes	Yes	No

is impossible. In Markov logic, formulas are soft constraints: a world that violates a formula is less probable than one that satisfies it, other things being equal, but not impossible. The weight of a formula represents its strength as a constraint. Finite first-order logic is the limit of Markov logic when all weights tend to infinity. Markov logic allows an existing first-order KB to be transformed into a probabilistic model simply by assigning weights to the formulas, manually or by learning them from data. It allows multiple KBs to be merged without resolving their inconsistencies, and obviates the need to exhaustively specify the conditions under which a formula can be applied. On the statistical side, it allows very complex models to be represented very compactly; in particular, it provides an elegant language for expressing non-i.i.d. models (i.e., models where data points are not assumed independent and identically distributed). It also facilitates the incorporation of rich domain knowledge, reducing reliance on purely empirical learning.

Markov logic builds on previous developments in knowledge-based model construction and statistical relational learning, but goes beyond them in combining first-order logic and graphical models without restrictions. Unlike previous representations, it is supported by a full range of learning and inference algorithms, in each case combining logical and statistical elements. Because of its generality, it provides a natural framework for integrating the logical and statistical approaches in each field. For example, DCGs and PCFGs are both special cases of Markov logic, and classical planning and MDPs can both be elegantly formulated using Markov logic and decision theory. With Markov logic, combining classical planning and MDPs, or DCGs and PCFGs, does not require new algorithms; the existing general-purpose inference and learning facilities can be directly applied.

AI can be roughly divided into "foundational" and "application" areas. Foundational areas include knowledge representation, automated reasoning, probabilistic models, and machine learning. Application areas include planning, vision, robotics, speech, natural language processing, and multi-agent systems. An interface layer for AI must provide the former, and serve the latter. We have developed the Alchemy system as an open-source embodiment of Markov logic and implementation of algorithms for it [60]. Alchemy seamlessly combines first-order knowledge representation, model checking, probabilistic inference, inductive logic programming, and generative/discriminative parameter learning. Table 1.3 compares Alchemy with Prolog and BUGS, and shows that it provides a critical mass of capabilities not previously available.

For researchers and practitioners in application areas, Alchemy offers a large reduction in the effort required to assemble a state-of-the-art solution, and to extend it beyond the state of the art.

The representation, inference and learning components required for each subtask, both logical and probabilistic, no longer need to be built or patched together piece by piece; Alchemy provides them, and the solution is built simply by writing formulas in Markov logic. A few lines of Alchemy suffice to build state-of-the-art systems for applications like collective classification, link prediction, entity resolution, information extraction, ontology mapping, and others. Because each of these pieces is now simple to implement, combining them into larger systems becomes straightforward, and is no longer a major engineering challenge. For example, we are currently beginning to build a complete natural language processing system in Alchemy, which aims to provide the functionality of current systems in one to two orders of magnitude fewer lines of code. Most significantly, Alchemy facilitates extending NLP systems beyond the current state of the art, for example by integrating probabilities into semantic analysis.

One of our goals with Alchemy is to support the growth of a repository of reusable knowledge bases in Markov logic, akin to the shareware repositories available today, and building on the traditional knowledge bases already available. Given such a repository, the first step of an application project becomes the selection of relevant knowledge. This may be used as is or manually refined. A new knowledge base is initiated by writing down plausible hypotheses about the new domain. This is followed by induction from data of new knowledge for the task. The formulas and weights of the supporting knowledge bases may also be adjusted based on data from the new task. New knowledge is added by noting and correcting the failures of the induced and refined KBs, and the process repeats. Over time, new knowledge is gradually accumulated, and existing knowledge is refined and specialized to different (sub)domains. Experience shows that neither knowledge engineering nor machine learning by itself is sufficient to reach human-level AI, and a simple two-stage solution of knowledge engineering followed by machine learning is also insufficient. What is needed is a fine-grained combination of the two, where each one bootstraps from the other, and at the end of each loop of bootstrapping the AI system's state of knowledge is more advanced than at the beginning. Alchemy supports this.

More broadly, a tool like Alchemy can help the focus of research shift from very specialized goals to higher-level ones. This is essential to speed progress in AI. As the field has grown, it has become atomized, but ultimately the pieces need to be brought back together. However, attempting to do this without an interface layer, by gluing together a large number of disparate pieces, rapidly turns into an engineering quagmire; systems become increasingly hard to build on for further progress, and eventually sheer complexity slows progress to a crawl. By keeping the pieces simpler and providing a uniform language for representing and combining them, even if at some cost in performance, an interface layer enables us to reach much farther before hitting the complexity wall. At that point, we have hopefully acquired the knowledge and insights to design the next higher-level interface layer, and in this way we can continue to make rapid progress.

Highly focused research is essential for progress, and often provides immediate real-world benefits in its own right. But these benefits will be dwarfed by those obtained if AI reaches and surpasses human intelligence, and to contribute toward this, improvements in performance in the subtasks need to translate into improvements in the larger tasks. When the subtasks are pursued in isolation, there is no guarantee that this will happen, and in fact experience suggests that the tendency will be for the subtask solutions to evolve into local optima, which are best in isolation but not in combination. By increasing the granularity of the tasks that can be routinely attempted, platforms like Alchemy make this less likely, and help us reach human-level AI sooner.

1.4   OVERVIEW OF THE BOOK

In the remainder of this book, we will describe the Markov logic representation in greater detail, along with algorithms and applications.

In Chapter 2, we first provide basic background on first-order logic and probabilistic graphical models. We then define the Markov logic representation, building and unifying these two perspectives. We conclude by showing how Markov logic relates to some of the many other combinations of logic and probability that have been proposed in recent years.

Chapters 3 and 4 present state-of-the-art algorithms for reasoning and learning with Markov logic. These algorithms build on standard methods for first-order logic or graphical models, including satisfiability, Markov chain Monte Carlo, and belief propagation for inference; and inductive logic programming and convex optimization for learning. In many cases, these methods have been combined and extended to handle additional challenges introduced by the rich Markov logic representation.

Chapter 5 goes beyond the basic Markov logic representation to describe several extensions that increase its power or applicability to particular problems. In particular, we cover how Markov logic can be extended to continuous and infinite domains, combined with decision theory, and generalized to represent uncertain disjunctions and existential quantifiers. In addition to solving particular problems better, these extensions demonstrate that Markov logic can easily be adapted when necessary to explicitly support the features of new problems.

Chapter 6 is devoted to exploring applications of Markov logic to several real-world problems, including collective classification, link prediction, link-based clustering, entity resolution, information extraction, social network analysis, and robot mapping. Most datasets and models from this chapter can be found online at http://alchemy.cs.washington.edu.

We conclude in Chapter 7 with final thoughts and future directions. An appendix provides a brief introduction to Alchemy.

Sample course slides to accompany this book are available at http://www.cs.washington.edu/homes/pedrod/803/.

CHAPTER 2

Markov Logic

In this chapter, we provide a detailed description of the Markov logic representation. We begin by providing background on first-order logic and probabilistic graphical models and then show how Markov logic unifies and builds on these concepts. Finally, we compare Markov logic to other representations that combine probability and logic.

2.1 FIRST-ORDER LOGIC

A *first-order knowledge base (KB)* is a set of sentences or formulas in first-order logic [37]. Formulas are constructed using four types of symbols: constants, variables, functions, and predicates. Constant symbols represent objects in the domain of interest (e.g., people: Anna, Bob, Chris, etc.). Variable symbols range over the objects in the domain. Function symbols (e.g., MotherOf) represent mappings from tuples of objects to objects. Predicate symbols represent relations among objects in the domain (e.g., Friends) or attributes of objects (e.g., Smokes). An *interpretation* specifies which objects, functions and relations in the domain are represented by which symbols. Variables and constants may be *typed*, in which case variables range only over objects of the corresponding type, and constants can only represent objects of the corresponding type. For example, the variable x might range over people (e.g., Anna, Bob, etc.), and the constant C might represent a city (e.g., Seattle, Tokyo, etc.).

A *term* is any expression representing an object in the domain. It can be a constant, a variable, or a function applied to a tuple of terms. For example, Anna, x, and GreatestCommonDivisor(x, y) are terms. An *atomic formula* or *atom* is a predicate symbol applied to a tuple of terms (e.g., Friends(x, MotherOf(Anna))). Formulas are recursively constructed from atomic formulas using logical connectives and quantifiers. If F_1 and F_2 are formulas, the following are also formulas: $\neg F_1$ (negation), which is true iff F_1 is false; $F_1 \wedge F_2$ (conjunction), which is true iff both F_1 and F_2 are true; $F_1 \vee F_2$ (disjunction), which is true iff F_1 or F_2 is true; $F_1 \Rightarrow F_2$ (implication), which is true iff F_1 is false or F_2 is true; $F_1 \Leftrightarrow F_2$ (equivalence), which is true iff F_1 and F_2 have the same truth value; $\forall x\ F_1$ (universal quantification), which is true iff F_1 is true for every object x in the domain; and $\exists x\ F_1$ (existential quantification), which is true iff F_1 is true for at least one object x in the domain. Parentheses may be used to enforce precedence. A *positive literal* is an atomic formula; a *negative literal* is a negated atomic formula. The formulas in a KB are implicitly conjoined, and thus a KB can be viewed as a single large formula. A *ground term* is a term containing no variables. A *ground atom* or *ground predicate* is an atomic formula all of whose arguments are ground terms. A *possible world* (along with an interpretation) assigns a truth value to each possible ground atom.

A formula is *satisfiable* iff there exists at least one world in which it is true. The basic inference problem in first-order logic is to determine whether a knowledge base *KB entails* a formula F, i.e.,

Table 2.1: Example of a first-order knowledge base and MLN. Fr() is short for `Friends()`, Sm() for `Smokes()`, and Ca() for `Cancer()`.

English and First-Order Logic	Clausal Form	Weight
"Friends of friends are friends."		
$\forall x \forall y \forall z$ Fr(x, y) $\wedge$ Fr(y, z) $\Rightarrow$ Fr(x, z)	$\neg$Fr(x, y) $\vee$ $\neg$Fr(y, z) $\vee$ Fr(x, z)	0.7
"Smoking causes cancer."		
$\forall x$ Sm(x) $\Rightarrow$ Ca(x)	$\neg$Sm(x) $\vee$ Ca(x)	1.5
"If two people are friends and one smokes, then so does the other."		
$\forall x \forall y$ Fr(x, y) $\wedge$ Sm(x) $\Rightarrow$ Sm(y)	$\neg$Fr(x, y) $\vee$ $\neg$Sm(x) $\vee$ Sm(y)	1.1

if F is true in all worlds where KB is true (denoted by $KB \models F$). This is often done by *refutation*: KB entails F iff $KB \cup \neg F$ is unsatisfiable. (Thus, if a KB contains a contradiction, all formulas trivially follow from it, which makes painstaking knowledge engineering a necessity.) For automated inference, it is often convenient to convert formulas to a more regular form, typically *clausal form* (also known as *conjunctive normal form (CNF)*). A KB in clausal form is a conjunction of *clauses*, a clause being a disjunction of literals. Every KB in first-order logic can be converted to clausal form using a mechanical sequence of steps.[1] Clausal form is used in resolution, a sound and refutation-complete inference procedure for first-order logic [122].

Inference in first-order logic is only semidecidable. Because of this, knowledge bases are often constructed using a restricted subset of first-order logic with more desirable properties. The most widely used restriction is to *Horn clauses*, which are clauses containing at most one positive literal. The Prolog programming language is based on Horn clause logic [72]. Prolog programs can be learned from databases by searching for Horn clauses that (approximately) hold in the data; this is studied in the field of inductive logic programming (ILP) [65].

Table 2.1 shows a simple KB and its conversion to clausal form. Note that, while these formulas may be *typically* true in the real world, they are not *always* true. In most domains it is very difficult to come up with non-trivial formulas that are always true, and such formulas capture only a fraction of the relevant knowledge. Thus, despite its expressiveness, pure first-order logic has limited applicability to practical AI problems. Many *ad hoc* extensions to address this have been proposed. In the more limited case of propositional logic, the problem is well solved by probabilistic graphical models such as Markov networks, described in the next section. We will later show how to generalize these models to the first-order case.

---

[1]This conversion includes the removal of existential quantifiers by Skolemization, which is not sound in general. However, in finite domains an existentially quantified formula can simply be replaced by a disjunction of its groundings.

2.2   MARKOV NETWORKS

A *Markov network* (also known as *Markov random field*) is a model for the joint distribution of a set of variables $X = (X_1, X_2, \ldots, X_n) \in \mathcal{X}$ [99]. It is composed of an undirected graph G and a set of potential functions ϕ_k. The graph has a node for each variable, and the model has a potential function for each clique in the graph. A potential function is a non-negative real-valued function of the state of the corresponding clique. The joint distribution represented by a Markov network is given by

$$P(X = x) = \frac{1}{Z} \prod_k \phi_k(x_{\{k\}}) \tag{2.1}$$

where $x_{\{k\}}$ is the state of the kth clique (i.e., the state of the variables that appear in that clique). Z, known as the *partition function*, is given by $Z = \sum_{x \in \mathcal{X}} \prod_k \phi_k(x_{\{k\}})$. Markov networks are often conveniently represented as *log-linear models*, with each clique potential replaced by an exponentiated weighted sum of features of the state, leading to

$$P(X = x) = \frac{1}{Z} \exp\left(\sum_j w_j f_j(x) \right) \tag{2.2}$$

A feature may be any real-valued function of the state. Except where stated, this book will focus on binary features, $f_j(x) \in \{0, 1\}$. In the most direct translation from the potential-function form (Equation 2.1), there is one feature corresponding to each possible state $x_{\{k\}}$ of each clique, with its weight being $\log \phi_k(x_{\{k\}})$. This representation is exponential in the size of the cliques. However, we are free to specify a much smaller number of features (e.g., logical functions of the state of the clique), allowing for a more compact representation than the potential-function form, particularly when large cliques are present. Markov logic will take advantage of this.

Inference in Markov networks is #P-complete [123]. The most widely used method for approximate inference in Markov networks is Markov chain Monte Carlo (MCMC) [40], and in particular Gibbs sampling, which proceeds by sampling each variable in turn given its Markov blanket. (The Markov blanket of a node is the minimal set of nodes that renders it independent of the remaining network; in a Markov network, this is simply the node's neighbors in the graph.) Marginal probabilities are computed by counting over these samples; conditional probabilities are computed by running the Gibbs sampler with the conditioning variables clamped to their given values.

Another popular method for inference in Markov networks is belief propagation [156], a message-passing algorithm that performs exact inference on tree-structured Markov networks. When applied to graphs with loops, the results are approximate and the algorithm may not converge. Nonetheless, loopy belief propagation is more efficient than Gibbs sampling in many applications.

Maximum-likelihood or MAP estimates of Markov network weights cannot be computed in closed form but, because the log-likelihood is a concave function of the weights, they can be found efficiently (modulo inference) using standard gradient-based or quasi-Newton optimization methods [95]. Another alternative is iterative scaling [24]. Features can also be learned from data, for example by greedily constructing conjunctions of atomic features [24].

2.3   MARKOV LOGIC

A first-order KB can be seen as a set of hard constraints on the set of possible worlds: if a world violates even one formula, it has zero probability. The basic idea in MLNs is to soften these constraints: when a world violates one formula in the KB it is less probable, but not impossible. The fewer formulas a world violates, the more probable it is. Each formula has an associated weight that reflects how strong a constraint it is: the higher the weight, the greater the difference in log probability between a world that satisfies the formula and one that does not, other things being equal.

Definition 2.1.   A *Markov logic network* L is a set of pairs (F_i, w_i), where F_i is a formula in first-order logic and w_i is a real number. Together with a finite set of constants $C = \{c_1, c_2, \ldots, c_{|C|}\}$, it defines a Markov network $M_{L,C}$ (Equations 2.1 and 2.2) as follows:

1. $M_{L,C}$ contains one binary node for each possible grounding of each predicate appearing in L. The value of the node is 1 if the ground predicate is true, and 0 otherwise.

2. $M_{L,C}$ contains one feature for each possible grounding of each formula F_i in L. The value of this feature is 1 if the ground formula is true, and 0 otherwise. The weight of the feature is the w_i associated with F_i in L.

The syntax of the formulas in an MLN is the standard syntax of first-order logic [37]. Free (unquantified) variables are treated as universally quantified at the outermost level of the formula. In this book, we will often assume that the set of formulas is in function-free clausal form for convenience, but our methods can be applied to other MLNs as well.

An MLN can be viewed as a *template* for constructing Markov networks. Given different sets of constants, it will produce different networks, and these may be of widely varying size, but all will have certain regularities in structure and parameters, given by the MLN (e.g., all groundings of the same formula will have the same weight). We call each of these networks a *ground Markov network* to distinguish it from the first-order MLN. From Definition 2.1 and Equations 2.1 and 2.2, the probability distribution over possible worlds x specified by the ground Markov network $M_{L,C}$ is given by

$$P(X=x) = \frac{1}{Z} \exp\left(\sum_i w_i n_i(x)\right) = \frac{1}{Z} \prod_i \phi_i(x_{\{i\}})^{n_i(x)} \tag{2.3}$$

where $n_i(x)$ is the number of true groundings of F_i in x, $x_{\{i\}}$ is the state (truth values) of the predicates appearing in F_i, and $\phi_i(x_{\{i\}}) = e^{w_i}$. Note that, although we defined MLNs as log-linear models, they could equally well be defined as products of potential functions, as the second equality above shows. This will be the most convenient approach in domains with a mixture of hard and

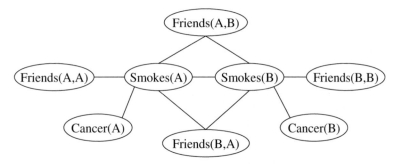

Figure 2.1: Ground Markov network obtained by applying the last two formulas in Table 2.1 to the constants Anna(A) and Bob(B).

soft constraints (i.e., where some formulas hold with certainty, leading to zero probabilities for some worlds).

The graphical structure of $M_{L,C}$ follows from Definition 2.1: there is an edge between two nodes of $M_{L,C}$ iff the corresponding ground predicates appear together in at least one grounding of one formula in L. Thus, the predicates in each ground formula form a (not necessarily maximal) clique in $M_{L,C}$. Figure 2.1 shows the graph of the ground Markov network defined by the last two formulas in Table 2.1 and the constants Anna and Bob. Each node in this graph is a ground predicate (e.g., Friends(Anna, Bob)). The graph contains an arc between each pair of predicates that appear together in some grounding of one of the formulas. $M_{L,C}$ can now be used to infer the probability that Anna and Bob are friends given their smoking habits, the probability that Bob has cancer given his friendship with Anna and whether she has cancer, etc.

Each state of $M_{L,C}$ represents a possible world. A possible world is a set of objects, a set of functions (mappings from tuples of objects to objects), and a set of relations that hold between those objects; together with an interpretation, they determine the truth value of each ground predicate. The following assumptions ensure that the set of possible worlds for (L, C) is finite, and that $M_{L,C}$ represents a unique, well-defined probability distribution over those worlds, irrespective of the interpretation and domain. These assumptions are quite reasonable in most practical applications, and greatly simplify the use of MLNs. For the remaining cases, we discuss below the extent to which each one can be relaxed.

Assumption 2.2.   Unique names. *Different constants refer to different objects [37].*

Assumption 2.3.   Domain closure. *The only objects in the domain are those representable using the constant and function symbols in (L, C) [37].*

Table 2.2: Construction of all groundings of a first-order formula under Assumptions 2.2–2.4.

---

function Ground(F)
   input: F, a formula in first-order logic
   output: G_F, a set of ground formulas
for each existentially quantified subformula $\exists x \; S(x)$ in F
   $F \leftarrow F$ with $\exists x \; S(x)$ replaced by $S(c_1) \vee S(c_2) \vee \ldots \vee S(c_{|C|})$,
     where $S(c_i)$ is $S(x)$ with x replaced by c_i
$G_F \leftarrow \{F\}$
for each universally quantified variable x
   for each formula $F_j(x)$ in G_F
     $G_F \leftarrow (G_F \setminus F_j(x)) \cup \{F_j(c_1), F_j(c_2), \ldots, F_j(c_{|C|})\}$,
       where $F_j(c_i)$ is $F_j(x)$ with x replaced by c_i
for each formula $F_j \in G_F$
   repeat
     for each function $f(a_1, a_2, \ldots)$ all of whose arguments are constants
       $F_j \leftarrow F_j$ with $f(a_1, a_2, \ldots)$ replaced by c, where $c = f(a_1, a_2, \ldots)$
   until F_j contains no functions
return G_F

Assumption 2.4. Known functions. *For each function appearing in L, the value of that function applied to every possible tuple of arguments is known, and is an element of C.*

This last assumption allows us to replace functions by their values when grounding formulas. Thus the only ground predicates that need to be considered are those having constants as arguments. The infinite number of terms constructible from all functions and constants in (L, C) (the "Herbrand universe" of (L, C)) can be ignored, because each of those terms corresponds to a known constant in C, and predicates involving them are already represented as the predicates involving the corresponding constants. The possible groundings of a predicate in Definition 2.1 are thus obtained simply by replacing each variable in the predicate with each constant in C, and replacing each function term in the predicate by the corresponding constant. Table 2.2 shows how the groundings of a formula are obtained given Assumptions 2.2–2.4.

Assumption 2.2 (unique names) can be removed by introducing the equality predicate (Equals(x, y), or x = y for short) and adding the necessary axioms to the MLN: equality is reflexive, symmetric and transitive; for each unary predicate P, $\forall x \forall y \; x = y \Rightarrow (P(x) \Leftrightarrow P(y))$; and similarly for higher-order predicates and functions [37]. The resulting MLN will have a node for each pair of constants, whose value is 1 if the constants represent the same object and 0 otherwise; these nodes will be connected to each other and to the rest of the network by arcs representing the axioms above.

Note that this allows us to make probabilistic inferences about the equality of two constants. The example in Section 6.3 successfully uses this as the basis of an approach to entity resolution.

If the number u of unknown objects is known, Assumption 2.3 (domain closure) can be removed simply by introducing u arbitrary new constants. If u is unknown but finite, Assumption 2.3 can be removed by introducing a distribution over u, grounding the MLN with each number of unknown objects, and computing the probability of a formula F as $P(F) = \sum_{u=0}^{u_{max}} P(u)P(F|M_{L,C}^u)$, where $M_{L,C}^u$ is the ground MLN with u unknown objects. Markov logic can also be applied to infinite domains; details are in Section 5.2.

Let $H_{L,C}$ be the set of all ground terms constructible from the function symbols in L and the constants in L and C (the "Herbrand universe" of (L, C)). Assumption 2.4 (known functions) can be removed by treating each element of $H_{L,C}$ as an additional constant and applying the same procedure used to remove the unique names assumption. For example, with a function $G(x)$ and constants A and B, the MLN will now contain nodes for $G(A) = A$, $G(A) = B$, etc. This leads to an infinite number of new constants, requiring the corresponding extension of MLNs. However, if we restrict the level of nesting to some maximum, the resulting MLN is still finite.

To summarize, Assumptions 2.2–2.4 can be removed as long the domain is finite. Section 5.2 discusses how to extend MLNs to infinite domains. In the remainder of this book we proceed under Assumptions 2.2–2.4, except where noted.

A first-order KB can be transformed into an MLN simply by assigning a weight to each formula. For example, the formulas (or clauses) and weights in the last two columns of Table 2.1 constitute an MLN. According to this MLN, other things being equal, a world where n smokers don't have cancer is $e^{1.5n}$ times less probable than a world where all smokers have cancer. Note that all the formulas in Table 2.1 are false in the real world as universally quantified logical statements, but capture useful information on friendships and smoking habits, when viewed as features of a Markov network. For example, it is well known that teenage friends tend to have similar smoking habits [73]. In fact, an MLN like the one in Table 2.1 succinctly represents a type of model that is a staple of social network analysis [149].

It is easy to see that MLNs subsume essentially all propositional probabilistic models, as detailed below.

Theorem 2.5. *Every probability distribution over discrete or finite-precision numeric variables can be represented as a Markov logic network.*

Proof. Consider first the case of Boolean variables $(X_1, X_2, \ldots, X_n)$. Define a predicate of zero arity R_h for each variable X_h, and include in the MLN L a formula for each possible state of $(X_1, X_2, \ldots, X_n)$. This formula is a conjunction of n literals, with the hth literal being $R_h()$ if X_h is true in the state, and $\neg R_h()$ otherwise. The formula's weight is $\log P(X_1, X_2, \ldots, X_n)$. (If some states have zero probability, use instead the product form (see Equation 2.3), with $\phi_i()$ equal to the probability of the ith state.) Since all predicates in L have zero arity, L defines the same Markov network $M_{L,C}$ irrespective of C, with one node for each variable X_h. For any state, the corresponding formula is true and all others are false, and thus Equation 2.3 represents the original distribution (notice that $Z = 1$). The generalization to arbitrary discrete variables is straightforward, by defining a zero-arity predicate for each value of each variable. Similarly for finite-precision numeric variables, by noting that they can be represented as Boolean vectors.   □

Of course, compact factored models like Markov networks and Bayesian networks can still be represented compactly by MLNs, by defining formulas for the corresponding factors (arbitrary features in Markov networks, and states of a node and its parents in Bayesian networks).[2]

First-order logic (with Assumptions 2.2–2.4 above) is the special case of MLNs obtained when all weights are equal and tend to infinity, as described below.

Theorem 2.6.   *Let KB be a satisfiable knowledge base, L be the MLN obtained by assigning weight w to every formula in KB, C be the set of constants appearing in KB, $P_w(x)$ be the probability assigned to a (set of) possible world(s) x by $M_{L,C}$, $\mathcal{X}_{KB}$ be the set of worlds that satisfy KB, and F be an arbitrary formula in first-order logic. Then:*

1. *$\forall x \in \mathcal{X}_{KB} \ \lim_{w\to\infty} P_w(x) = |\mathcal{X}_{KB}|^{-1}$*
   $\forall x \notin \mathcal{X}_{KB} \ \lim_{w\to\infty} P_w(x) = 0$

2. *For all F, $KB \models F$ iff $\lim_{w\to\infty} P_w(F) = 1$.*

Proof. Let k be the number of ground formulas in $M_{L,C}$. By Equation 2.3, if $x \in \mathcal{X}_{KB}$ then $P_w(x) = e^{kw}/Z$, and if $x \notin \mathcal{X}_{KB}$ then $P_w(x) \le e^{(k-1)w}/Z$. Thus all $x \in \mathcal{X}_{KB}$ are equiprobable and $\lim_{w\to\infty} P(\mathcal{X} \setminus \mathcal{X}_{KB})/P(\mathcal{X}_{KB}) \le \lim_{w\to\infty}(|\mathcal{X} \setminus \mathcal{X}_{KB}|/|\mathcal{X}_{KB}|)e^{-w} = 0$, proving Part 1. By definition of entailment, $KB \models F$ iff every world that satisfies KB also satisfies F. Therefore, letting $\mathcal{X}_F$ be the set of worlds that satisfy F, if $KB \models F$ then $\mathcal{X}_{KB} \subseteq \mathcal{X}_F$ and $P_w(F) = \sum_{x\in\mathcal{X}_F} P_w(x) \ge P_w(\mathcal{X}_{KB})$. Since, from Part 1, $\lim_{w\to\infty} P_w(\mathcal{X}_{KB}) = 1$, this implies that if $KB \models F$ then $P_w(F) = 1$. The inverse direction of Part 2 is proved by noting that if $P_w(F) = 1$ then every world with non-zero probability must satisfy F, and this includes every world in $\mathcal{X}_{KB}$.   □

In other words, in the limit of all equal infinite weights, the MLN represents a uniform distribution over the worlds that satisfy the KB, and all entailment queries can be answered by computing the probability of the query formula and checking whether it is 1. Even when weights are finite, first-order logic is "embedded" in MLNs in the following sense. Assume without loss of generality that all weights are non-negative. (A formula with a negative weight w can be replaced by its negation with weight $-w$.) If the knowledge base composed of the formulas in an MLN L (negated, if their weight is negative) is satisfiable, then, for any C, the satisfying assignments are the modes of the distribution represented by $M_{L,C}$. This is because the modes are the worlds x with maximum $\sum_i w_i n_i(x)$ (see Equation 2.3), and this expression is maximized when all groundings of

---

[2]While some conditional independence structures can be compactly represented with directed graphs but not with undirected ones, they still lead to compact models in the form of Equation 2.3 (i.e., as products of potential functions).

all formulas are true (i.e., the KB is satisfied). Unlike an ordinary first-order KB, however, an MLN can produce useful results even when it contains contradictions. An MLN can also be obtained by merging several KBs, even if they are partly incompatible. This is potentially useful in areas like the Semantic Web [5] and mass collaboration [116].

It is interesting to see a simple example of how MLNs generalize first-order logic. Consider an MLN containing the single formula $\forall x\, R(x) \Rightarrow S(x)$ with weight w, and $C = \{A\}$. This leads to four possible worlds: $\{\neg R(A), \neg S(A)\}, \{\neg R(A), S(A)\}, \{R(A), \neg S(A)\}$, and $\{R(A), S(A)\}$. From Equation 2.3 we obtain that $P(\{R(A), \neg S(A)\}) = 1/(3e^w + 1)$ and the probability of each of the other three worlds is $e^w/(3e^w + 1)$. (The denominator is the partition function Z; see Section 2.2.) Thus, if $w > 0$, the effect of the MLN is to make the world that is inconsistent with $\forall x\, R(x) \Rightarrow S(x)$ less likely than the other three. From the probabilities above we obtain that $P(S(A)|R(A)) = 1/(1 + e^{-w})$. When $w \to \infty$, $P(S(A)|R(A)) \to 1$, recovering the logical entailment.

A first-order KB partitions the set of possible worlds into two subsets: those that satisfy the KB and those that do not. An MLN has many more degrees of freedom: it can partition the set of possible worlds into many more subsets, and assign a different probability to each. How to use this freedom is a key decision for both knowledge engineering and learning. At one extreme, the MLN can add little to logic, treating the whole knowledge base as a single formula, and assigning one probability to the worlds that satisfy it and another to the worlds that do not. At the other extreme, each formula in the KB can be converted into clausal form, and a weight associated with each clause.[3] The more finely divided into subformulas a KB is, the more gradual the drop-off in probability as a world violates more of those subformulas, and the greater the flexibility in specifying distributions over worlds. From a knowledge engineering point of view, the decision about which formulas constitute indivisible constraints should reflect domain knowledge and the goals of modeling. From a learning point of view, dividing the KB into more formulas increases the number of parameters, with the corresponding tradeoff in bias and variance.

It is also interesting to see an example of how MLNs generalize commonly-used statistical models. One of the most widely used models for classification is logistic regression. Logistic regression predicts the probability that an example with features $f = (f_1, \ldots, f_i, \ldots)$ is of class c according to the equation:

$$\log\left(\frac{P(C = 1|F = f)}{P(C = 0|F = f)}\right) = a + \sum_{i=1}^{n} b_i f_i$$

This can be implemented as an MLN using a unit clause for the class, $C(x)$, with weight a, and a formula of the form $F_i(x) \wedge C(x)$ for each feature, with weight b_i. This yields the distribution

$$P(C = c, F = f) = \frac{1}{Z} \exp\left(ac + \sum_i b_i f_i c\right)$$

---

[3]This conversion can be done in the standard way [37], except that, instead of introducing Skolem functions, existentially quantified formulas should be replaced by disjunctions, as in Table 2.2.

resulting in

$$\frac{P(C = 1|F = f)}{P(C = 0|F = f)} = \frac{\exp\left(a + \sum_i b_i f_i\right)}{\exp(0)} = \exp\left(a + \sum_i^n b_i f_i\right)$$

as desired.

In practice, we have found it useful to add each predicate to the MLN as a unit clause. In other words, for each predicate $R(x_1, x_2, \dots)$ appearing in the MLN, we add the formula $\forall x_1, x_2, \dots R(x_1, x_2, \dots)$ with some weight w_R. The weight of a unit clause can (roughly speaking) capture the marginal distribution of the corresponding predicate, leaving the weights of the non-unit clauses free to model only dependencies between predicates.

When manually constructing an MLN or interpreting a learned one, it is useful to have an intuitive understanding of the weights. Consider a ground formula F with weight w. All other things being equal, a world where F is true is e^w times as likely as a world where F is false. Let U_t and U_f be the number of possible worlds in which F is true and false, respectively. If F is independent from all other ground formulas, then its probability is given by the following function:

$$P(F) = \frac{1}{1 + \frac{U_f}{U_t}e^{-w}}$$

Solving for w:

$$w = \log\frac{P(F)}{P(\neg F)} - \log\frac{U_t}{U_f}$$

Therefore, w can be interpreted as the difference between the log odds of F according to the MLN and according to the uniform distribution. However, if F shares atoms with other formulas, as will typically be the case, it may not be possible to keep those formulas' truth values unchanged while reversing F's. In this case, there is no longer a one-to-one correspondence between weights and probabilities of formulas.[4] Nevertheless, the probabilities of all formulas collectively determine all weights if we view them as constraints on a maximum entropy distribution, or treat them as empirical probabilities and learn the maximum likelihood weights (the two are equivalent) [24]. Thus a good way to set the weights of an MLN is to write down the probability with which each formula should hold, treat these as empirical frequencies, and learn the weights from them using the algorithms in Section 4.1. Conversely, the weights in a learned MLN can be viewed as collectively encoding the empirical formula probabilities.

The size of ground Markov networks can be vastly reduced by having typed constants and variables, and only grounding variables to constants of the same type. However, even in this case the size of the network may still be extremely large. Fortunately, there are a number of ways to further reduce this size, as we will see in Chapter 3.

---

[4]This is an unavoidable side-effect of the power and flexibility of Markov networks. In Bayesian networks, parameters are probabilities, but at the cost of greatly restricting the ways in which the distribution may be factored. In particular, potential functions must be conditional probabilities, and the directed graph must have no cycles. The latter condition is particularly troublesome to enforce in relational extensions [141].

2.4   RELATION TO OTHER APPROACHES

There is a very large literature relating logic and probability; here we will focus only on the approaches most closely related to Markov logic.

EARLY WORK

Attempts to combine logic and probability in AI date back to at least Nilsson [94]. Bacchus [1], Halpern [43] and coworkers (e.g., [2]) studied the problem in detail from a theoretical standpoint. They made a distinction between statistical statements (e.g., "65% of the students in our department are undergraduate") and statements about possible worlds (e.g., "The probability that Anna is an undergraduate is 65%"), and provided methods for computing the latter from the former. In their approach, a KB did not specify a complete and unique distribution over possible worlds, leaving its status as a probabilistic model unclear. MLNs overcome this limitation by viewing KBs as Markov network templates.

Paskin [97] extended the work of Bacchus *et al.* [2] by associating a probability with each first-order formula, and taking the maximum entropy distribution compatible with those probabilities. This representation was still quite brittle, with a world that violates a single grounding of a universally quantified formula being considered as unlikely as a world that violates all of them. In contrast, in MLNs a rule like $\forall x\ \texttt{Smokes(x)} \Rightarrow \texttt{Cancer(x)}$ causes the probability of a world to decrease smoothly as the number of cancer-free smokers in it increases.

KNOWLEDGE-BASED MODEL CONSTRUCTION

Knowledge-based model construction (KBMC) is a combination of logic programming and Bayesian networks [151, 93, 55]. As in MLNs, nodes in KBMC represent ground predicates. Given a Horn KB, KBMC answers a query by finding all possible backward-chaining proofs of the query and evidence predicates from each other, constructing a Bayesian network over the ground predicates in the proofs, and performing inference over this network. The parents of a predicate node in the network are deterministic AND nodes representing the bodies of the clauses that have that node as the head. The conditional probability of the node given these is specified by a combination function (e.g., noisy OR or logistic regression). MLNs have several advantages compared to KBMC: they allow arbitrary formulas (not just Horn clauses) and inference in any direction, they sidestep the thorny problem of avoiding cycles in the Bayesian networks constructed by KBMC, and they do not require the introduction of *ad hoc* combination functions for clauses with the same consequent.

A KBMC model can be translated into an MLN by writing down a set of formulas for each first-order predicate $P_k(...)$ in the domain. Each formula is a conjunction containing $P_k(...)$ and one literal per parent of $P_k(...)$ (i.e., per first-order predicate appearing in a Horn clause having $P_k(...)$ as the consequent). A subset of these literals are negated; there is one formula for each possible combination of positive and negative literals. The weight of the formula is $w = \log[p/(1-p)]$, where p is the conditional probability of the child predicate when the corresponding conjunction of parent literals is true, according to the combination function used. If the combination function is logistic regression, it can be represented using only a linear number of formulas, taking advantage of the fact that a logistic regression model is a (conditional) Markov network with a binary clique between each predictor and the response. Noisy OR can similarly be represented with a linear number of parents.

OTHER LOGIC PROGRAMMING APPROACHES

Stochastic logic programs (SLPs) [87, 17] are a combination of logic programming and log-linear models. Puech and Muggleton [111] showed that SLPs are a special case of KBMC, and thus they can be converted into MLNs in the same way. Like MLNs, SLPs have one coefficient per clause, but they represent distributions over Prolog proof trees rather than over predicates; the latter have to be obtained by marginalization. Similar remarks apply to a number of other representations that are essentially equivalent to SLPs, like independent choice logic [103] and PRISM [129].

MACCENT [23] is a system that learns log-linear models with first-order features; each feature is a conjunction of a class and a Prolog query (clause with empty head). A key difference between MACCENT and MLNs is that MACCENT is a classification system (i.e., it predicts the conditional distribution of an object's class given its properties), while an MLN represents the full joint distribution of a set of predicates. Like any probability estimation approach, MLNs can be used for classification simply by issuing the appropriate conditional queries.[5] In particular, a MACCENT model can be converted into an MLN simply by defining a class predicate, adding the corresponding features and their weights to the MLN, and adding a formula with infinite weight stating that each object must have exactly one class. (This fails to model the marginal distribution of the non-class predicates, which is not a problem if only classification queries will be issued.) MACCENT can make use of deterministic background knowledge in the form of Prolog clauses; these can be added to the MLN as formulas with infinite weight. In addition, MLNs allow uncertain background knowledge (via formulas with finite weights). As we demonstrate in Section 6.1, MLNs can be used for collective classification, where the classes of different objects can depend on each other; MACCENT, which requires that each object be represented in a separate Prolog knowledge base, does not have this capability.

Constraint logic programming (CLP) is an extension of logic programming where variables are constrained instead of being bound to specific values during inference [64]. Probabilistic CLP generalizes SLPs to CLP [121], and CLP($\mathcal{B}N$) combines CLP with Bayesian networks [128]. Unlike in MLNs, constraints in CLP($\mathcal{B}N$) are hard (i.e., they cannot be violated; rather, they define the form of the probability distribution).

PROBABILISTIC RELATIONAL MODELS

Probabilistic relational models (PRMs) [36] are a combination of frame-based systems and Bayesian networks. PRMs can be converted into MLNs by defining a predicate $S(x, v)$ for each (propositional or relational) attribute of each class, where $S(x, v)$ means "The value of attribute S in object x is v."

---

[5]Conversely, joint distributions can be built up from classifiers (e.g., [44]), but this would be a significant extension of MACCENT.

A PRM is then translated into an MLN by writing down a formula for each line of each (class-level) conditional probability table (CPT) and value of the child attribute. The formula is a conjunction of literals stating the parent values and a literal stating the child value, and its weight is the logarithm of $P(x|Parents(x))$, the corresponding entry in the CPT. In addition, the MLN contains formulas with infinite weight stating that each attribute must take exactly one value. This approach handles all types of uncertainty in PRMs (attribute, reference and existence uncertainty).

As Taskar *et al.* [141] point out, the need to avoid cycles in PRMs causes significant representational and computational difficulties. Inference in PRMs is done by creating the complete ground network, which limits their scalability. PRMs require specifying a complete conditional model for each attribute of each class, which in large complex domains can be quite burdensome. In contrast, MLNs create a complete joint distribution from whatever number of first-order features the user chooses to specify.

RELATIONAL MARKOV NETWORKS
Relational Markov networks (RMNs) use database queries as clique templates, and have a feature for each state of a clique [141]. MLNs generalize RMNs by providing a more powerful language for constructing features (first-order logic instead of conjunctive queries), and by allowing uncertainty over arbitrary relations (not just attributes of individual objects). RMNs are exponential in clique size, while MLNs allow the user (or learner) to determine the number of features, making it possible to scale to much larger clique sizes.

STRUCTURAL LOGISTIC REGRESSION
In structural logistic regression (SLR) [109], the predictors are the output of SQL queries over the input data. Just as a logistic regression model is a discriminatively-trained Markov network, an SLR model is a discriminatively-trained MLN.[6]

RELATIONAL DEPENDENCY NETWORKS
In a relational dependency network (RDN), each node's probability conditioned on its Markov blanket is given by a decision tree [90]. Every RDN has a corresponding MLN in the same way that every dependency network has a corresponding Markov network, given by the stationary distribution of a Gibbs sampler operating on it [44].

PLATES AND PROBABILISTIC ER MODELS
Large graphical models with repeated structure are often compactly represented using plates [14]. MLNs subsume plates as a representation language. In addition, they allow individuals and their relations to be explicitly represented (see [18]), and context-specific independencies to be compactly written down, instead of left implicit in the node models. More recently, Heckerman *et al.* [46] have proposed a language based on entity-relationship models that combines the features of plates and

---

[6]Use of SQL aggregates requires that their definitions be imported into the MLN.

PRMs; this language is a special case of MLNs in the same way that ER models are a special case of logic. Probabilistic ER models allow logical expressions as constraints on how ground networks are constructed, but the truth values of these expressions have to be known in advance; MLNs allow uncertainty over all logical expressions.

BLOG

Milch *et al.* [84] have proposed a language called BLOG, designed to avoid making the unique names and domain closure assumptions. A BLOG program specifies procedurally how to generate a possible world, and does not allow arbitrary first-order knowledge to be easily incorporated. Also, it only specifies the structure of the model, leaving the parameters to be specified by external calls. BLOG models are directed graphs and need to avoid cycles, which substantially complicates their design. We saw in the previous section how to remove the unique names and domain closure assumptions in MLNs. (When there are unknown objects of multiple types, a random variable for the number of each type is introduced.) Inference about an object's attributes, rather than those of its observations, can be done simply by having variables for objects as well as for their observations (e.g., for books as well as citations to them). To our knowledge, BLOG has not yet been evaluated on any real-world problems, and no learning or general-purpose efficient inference algorithms for it have been developed.

FURTHER READING

See Richardson and Domingos [117] for a somewhat expanded introduction to the Markov logic representation. See Getoor and Taskar [39] for details on other representations that combine logic and probability. For more background on first-order logic, see Genesereth and Nilsson [37] or other textbooks. For more background on Markov networks and other graphical models, see Koller and Friedman [61] or Pearl [99].

CHAPTER 3

Inference

Inference in Markov logic lets us reason probabilistically about complex relationships. Since an MLN acts as a template for a Markov network, we can always answer probabilistic queries using standard Markov network inference methods on the instantiated network. However, due to the size and complexity of the resulting network, this is often infeasible. Instead, the methods we discuss here combine probabilistic methods with ideas from logical inference, including satisfiability and resolution. This leads to efficient methods that take full advantage of the logical structure.

We consider two basic types of inference: finding the most likely state of the world consistent with some evidence, and computing arbitrary conditional probabilities. We then discuss two approaches to making inference more tractable on large, relational problems: lazy inference, in which only the groundings that deviate from a "default" value need to be instantiated; and lifted inference, in which we group indistinguishable atoms together and treat them as a single unit during inference.

3.1 INFERRING THE MOST PROBABLE EXPLANATION

A basic inference task is finding the most probable state of the world y given some evidence x, where x is a set of literals. (This is known as MAP inference in the Markov network literature, and MPE inference in the Bayesian network literature.[1]) For Markov logic, this is formally defined as follows:

$$
\begin{aligned}
\arg\max_{y} P(y|x) &= \arg\max_{y} \frac{1}{Z_x} \exp\left(\sum_i w_i n_i(x, y)\right) \\
&= \arg\max_{y} \sum_i w_i n_i(x, y)
\end{aligned}
\tag{3.1}
$$

The first equality is due to Equation 2.3, which defines of the probability of a possible world. The normalization constant is written as Z_x to reflect the fact that we are only normalizing over possible worlds consistent with x. In the second equality, we remove Z_x since, being constant, it does not affect the arg max operation. We can also remove the exponentiation because it is a monotonic function. Therefore, the MPE problem in Markov logic reduces to finding the truth assignment that maximizes the sum of weights of satisfied clauses.

This can be done using any weighted satisfiability solver, and (remarkably) need not be more expensive than standard logical inference by model checking. (In fact, it can be faster, if some hard constraints are softened.) The problem is NP-hard in general, but effective solvers exist, both exact and approximate. The most commonly used approximate solver is MaxWalkSAT, a weighted variant

---

[1] The term "MAP inference" is sometimes used to refer to finding the most probable configuration of a set of query variables, given some evidence. The necessity of summing out all non-query, non-evidence variables makes this a harder inference problem than the one we consider here, in which y is the *complete* state of the world.

Table 3.1: MaxWalkSAT algorithm for MPE inference.

---

function MaxWalkSAT(L, m_t, m_f, *target*, p)
  inputs: L, a set of weighted clauses
        m_t, the maximum number of tries
        m_f, the maximum number of flips
        target, target solution cost
        p, probability of taking a random step
  output: *soln*, best variable assignment found
vars ← variables in L
for i ← 1 **to** m_t
  soln ← a random truth assignment to *vars*
  cost ← sum of weights of unsatisfied clauses in *soln*
  for i ← 1 **to** m_f
    if *cost* ≤ *target*
      return "Success, solution is", *soln*
    c ← a randomly chosen unsatisfied clause
    if Uniform(0,1) < p
      v_f ← a randomly chosen variable from c
    else
      for each variable v in c
        compute DeltaCost(v)
      v_f ← v with lowest DeltaCost(v)
    soln ← *soln* with v_f flipped
    cost ← *cost* + DeltaCost(v_f)
return "Failure, best assignment is", best *soln* found

---

of the WalkSAT local-search satisfiability solver, which can solve hard problems with hundreds of thousands of variables in minutes [53]. MaxWalkSAT performs this stochastic search by repeatedly picking an unsatisfied clause at random and flipping the truth value of one of the atoms in it. With a certain probability, the atom is chosen randomly; otherwise, the atom is chosen to maximize the sum of satisfied clause weights when flipped. This combination of random and greedy steps allows MaxWalkSAT to avoid getting stuck in local optima while searching. Pseudocode for MaxWalkSAT is shown in Table 3.1. DeltaCost(v) computes the change in the sum of weights of unsatisfied clauses that results from flipping variable v in the current solution. Uniform(0,1) returns a uniform deviate from the interval [0, 1].

3.2 COMPUTING CONDITIONAL PROBABILITIES

MLNs can answer arbitrary queries of the form "What is the probability that formula F_1 holds given that formula F_2 does?" If F_1 and F_2 are two formulas in first-order logic, C is a finite set of constants including any constants that appear in F_1 or F_2, and L is an MLN, then

$$
\begin{aligned}
P(F_1|F_2, L, C) &= P(F_1|F_2, M_{L,C}) \\
&= \frac{P(F_1 \wedge F_2|M_{L,C})}{P(F_2|M_{L,C})} \\
&= \frac{\sum_{x \in \mathcal{X}_{F_1} \cap \mathcal{X}_{F_2}} P(X=x|M_{L,C})}{\sum_{x \in \mathcal{X}_{F_2}} P(X=x|M_{L,C})}
\end{aligned} \tag{3.2}
$$

where $\mathcal{X}_{F_i}$ is the set of worlds where F_i holds, $M_{L,C}$ is the Markov network defined by L and C, and $P(X=x|M_{L,C})$ is given by Equation 2.3. Ordinary conditional queries in graphical models are the special case of Equation 3.2 where all predicates in F_1, F_2 and L are zero-arity and the formulas are conjunctions. The question of whether a knowledge base KB entails a formula F in first-order logic is the question of whether $P(F|L_{KB}, C_{KB,F}) = 1$, where L is the MLN obtained by assigning infinite weight to all the formulas in KB, and $C_{KB,F}$ is the set of all constants appearing in KB or F. The question is answered by computing $P(F|L_{KB}, C_{KB,F})$ by Equation 3.2, with F_2 = True.

Computing Equation 3.2 directly is intractable in all but the smallest domains. Since MLN inference subsumes probabilistic inference, which is #P-complete, and logical inference, which is NP-complete even in finite domains, no better results can be expected. However, many of the large number of techniques for efficient inference in either case are applicable to MLNs. Because MLNs allow fine-grained encoding of knowledge, including context-specific independences, inference in them may in some cases be more efficient than inference in an ordinary graphical model for the same domain. On the logic side, the probabilistic semantics of MLNs allows for approximate inference, with the corresponding potential gains in efficiency.

In principle, $P(F_1|F_2, L, C)$ can be approximated using an MCMC algorithm that rejects all moves to states where F_2 does not hold, and counts the number of samples in which F_1 holds. However, this may be too slow for arbitrary formulas. Instead, we focus on the case where F_2 is a conjunction of ground literals. While less general than Equation 3.2, this is the most frequent type of query in practice. In this scenario, further efficiency can be gained by applying a generalization of knowledge-based model construction [151]. The basic idea is to only construct the minimal subset of the ground network required to answer the query. This network is constructed by checking if the atoms that the query formula directly depends on are in the evidence. If they are, the construction is complete. Those that are not are added to the network, and we in turn check the atoms they depend on. This process is repeated until all relevant atoms have been retrieved. While in the worst case it yields no savings, in practice it can vastly reduce the time and memory required for inference.

Pseudocode for the network construction algorithm is shown in Table 3.2. See Figure 3.1 for an example of the resulting network. The size of the network returned may be further reduced, and the algorithm sped up, by noticing that any ground formula that is made true by the evidence can

Table 3.2: Network construction algorithm for inference in MLNs.

---

function ConstructNetwork(F_1, F_2, L, C)
    inputs: F_1, a set of ground predicates with unknown truth values (the "query")
           F_2, a set of ground predicates with known truth values (the "evidence")
           L, a Markov logic network
           C, a set of constants
    output: M, a ground Markov network
    calls: $MB(q)$, the Markov blanket of q in $M_{L,C}$
    $G \leftarrow F_1$
    while $F_1 \neq \emptyset$
        for all $q \in F_1$
           if $q \notin F_2$
               $F_1 \leftarrow F_1 \cup (MB(q) \setminus G)$
               $G \leftarrow G \cup MB(q)$
           $F_1 \leftarrow F_1 \setminus \{q\}$
    return M, the ground Markov network composed of all nodes in G, all arcs
    between them in $M_{L,C}$, and the features and weights on the corresponding cliques.

---

be ignored, and the corresponding arcs removed from the network. In the worst case, the network contains $O(|C|^a)$ nodes, where a is the largest predicate arity in the domain, but in practice, it may be much smaller.

Once the network has been constructed, we can apply any standard inference technique for Markov networks, such as Gibbs sampling, with the nodes in F_2 set to their respective values in F_2. The basic Gibbs step consists of sampling one ground predicate given its Markov blanket. The Markov blanket of a ground predicate is the set of ground predicates that appear in some grounding of a formula with it. The probability of a ground predicate X_l when its Markov blanket B_l is in state b_l is

$$P(X_l = x_l | B_l = b_l) \qquad (3.3)$$
$$= \frac{\exp(\sum_{f_i \in F_l} w_i f_i(X_l = x_l, B_l = b_l))}{\exp(\sum_{f_i \in F_l} w_i f_i(X_l = 0, B_l = b_l)) + \exp(\sum_{f_i \in F_l} w_i f_i(X_l = 1, B_l = b_l))}$$

where F_l is the set of ground formulas that X_l appears in, and $f_i(X_l = x_l, B_l = b_l)$ is the value (0 or 1) of the feature corresponding to the ith ground formula when $X_l = x_l$ and $B_l = b_l$. For sets of predicates of which exactly one is true in any given world (e.g., the possible values of an attribute), blocking can be used (i.e., one predicate is set to true and the others to false in one step, by sampling conditioned on their collective Markov blanket). The estimated probability of a

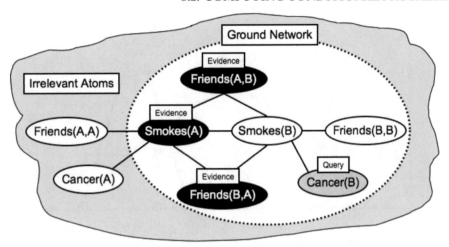

Figure 3.1: Ground network for computing $P(\text{Cancer}(B)|\text{Smokes}(A), \text{Friends}(A, B),$ $\text{Friends}(B, A))$, as built by the ConstructNetwork algorithm (Table 3.2). The constants and network used are the same as in Figure 2.1, but for this query and evidence the atoms Friends(A, A) and Cancer(A) are irrelevant and can be omitted from the network.

conjunction of ground literals is simply the fraction of samples in which the ground literals are true, after the Markov chain has converged.

HANDLING DETERMINISTIC DEPENDENCIES

One problem with using Gibbs sampling for inference in MLNs is that it breaks down in the presence of deterministic or near-deterministic dependencies (as do other probabilistic inference methods, e.g., belief propagation [156]). Deterministic dependencies break up the space of possible worlds into regions that are not reachable from each other, violating a basic requirement of MCMC. Near-deterministic dependencies greatly slow down inference, by creating regions of low probability that are very difficult to traverse. Running multiple chains with random starting points does not solve this problem, because it does not guarantee that each region will be sampled with frequency proportional to their probability, and there may be a very large number of regions.

The MC-SAT algorithm addresses this problem by combining MCMC with satisfiability testing [105]. MC-SAT is a *slice sampling* MCMC algorithm which uses a combination of satisfiability testing and simulated annealing to sample from the slice. The advantage of using a satisfiability solver (WalkSAT) is that it efficiently finds isolated modes in the distribution, and as a result the Markov chain mixes very rapidly. The slice sampling scheme ensures that detailed balance is (approximately) preserved.

MC-SAT is orders of magnitude faster than standard MCMC methods such as Gibbs sampling and simulated tempering, and is applicable to any model that can be expressed in Markov

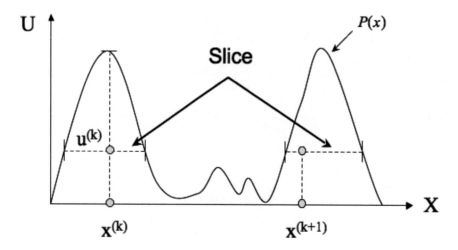

Figure 3.2: Visual illustration of the MCMC slice sampling method. Given the kth sample, $x^{(k)}$, we first sample the auxiliary variable $u^{(k)}$ uniformly from the range $[0, P(x^{(k)}]$. The next sample in the Markov chain, $x^{(k+1)}$, is sampled uniformly from all x with $P(x) \geq u^{(k)}$. This set of candidate x's is called the *slice*.

logic, including many standard models in statistical physics, vision, natural language processing, social network analysis, spatial statistics, etc.

Slice sampling [19] is an instance of a widely used approach in MCMC inference that introduces *auxiliary variables*, u, to capture the dependencies between observed variables, x. For example, to sample from $P(X\!=\!x) = (1/Z) \prod_k \phi_k(x_{\{k\}})$, we can define $P(X\!=\!x, U\!=\!u) = (1/Z) \prod_k I_{[0,\phi_k(x_{\{k\}})]}(u_k)$, where ϕ_k is the kth potential function, u_k is the kth auxiliary variable, $I_{[a,b]}(u_k) = 1$ if $a \leq u_k \leq b$, and $I_{[a,b]}(u_k) = 0$ otherwise. The marginal distribution of X under this joint is $P(X\!=\!x)$, so to sample from the original distribution it suffices to sample from $P(x, u)$ and ignore the u values. This can be done by alternately sampling $P(u|x)$ and $P(x|u)$. See Figure 3.2 for a graphical example of slice sampling.

$P(u_k|x)$ is uniform in $[0, \phi_k(x_{\{k\}})]$, and thus easy to sample from. The main challenge is to sample x given u, which is uniform among all $\mathcal{X}$ that satisfy $\phi_k(x_{\{k\}}) \geq u_k$ for all k. MC-SAT uses SampleSAT [150] to do this. In each sampling step, MC-SAT takes the set of all ground clauses satisfied by the current state of the world and constructs a subset, M, that must be satisfied by the next sampled state of the world. (For the moment we will assume that all clauses have positive weight.) Specifically, a satisfied ground clause is included in M with probability $1 - e^{-w}$, where w is the clause's weight. We then take as the next state a uniform sample from the set of states $SAT(M)$ that satisfy M. (Notice that $SAT(M)$ is never empty because it always contains at least the current state.) Table 3.3 gives pseudocode for MC-SAT. $\mathcal{U}_S$ is the uniform distribution over set S. At each step,

Table 3.3: Efficient MCMC inference algorithm for MLNs.

---

function MC-SAT(L, n)
   inputs: L, a set of weighted clauses $\{(w_j, c_j)\}$
         n, number of samples
   output: $\{x^{(1)}, \ldots, x^{(n)}\}$, set of n samples
  $x^{(0)} \leftarrow$ Satisfy(hard clauses in L)
  for $i \leftarrow 1$ **to** n
    $M \leftarrow \emptyset$
    for all $(w_k, c_k) \in L$ satisfied by $x^{(i-1)}$
      With probability $1 - e^{-w_k}$ add c_k to M
    Sample $x^{(i)} \sim \mathcal{U}_{SAT(M)}$

---

all hard clauses are selected with probability 1, and thus all sampled states satisfy them. Negative weights are handled by noting that a clause with weight $w < 0$ is equivalent to its negation with weight $-w$, and a clause's negation is the conjunction of the negations of all of its literals. Thus, instead of checking whether the clause is satisfied, we check whether its negation is satisfied; if it is, with probability $1 - e^w$ we select all of its negated literals, and with probability e^w we select none.

It can be shown that MC-SAT satisfies the MCMC criteria of detailed balance and ergodicity [105], assuming a perfect uniform sampler. In general, uniform sampling is #P-hard and SampleSAT [150] only yields approximately uniform samples. However, experiments show that MC-SAT is still able to produce very accurate probability estimates, and its performance is not very sensitive to the parameter settings of SampleSAT.

3.3   LAZY INFERENCE

One problem with the aforementioned approaches is that they require propositionalizing the domain (i.e., grounding all atoms and clauses in all possible ways), which consumes memory exponential in the arity of the clauses. Lazy inference methods [135, 108] overcome this by only grounding atoms and clauses as needed. This takes advantage of the sparseness of relational domains, where most atoms are false and most clauses are trivially satisfied. For example, in the domain of scientific research papers, most groundings of the atom Author(person, paper) are false, and most groundings of the clause Author(p1, paper) ∧ Author(p2, paper) ⇒ Coauthor(p1, p2) are trivially satisfied. With lazy inference, the memory cost does not scale with the number of possible clause groundings, but only with the number of groundings that have non-default values at some point in the inference.

We first describe a general approach for making inference algorithms lazy and then show how it can be applied to create lazy versions of MaxWalkSAT and MC-SAT. This method applies to any algorithm that works by repeatedly selecting a function or variable and recomputing its value depending only on a subset of the other variables/functions. These include a diverse set of algorithms for a variety of applications: WalkSAT, MaxWalkSAT, SampleSAT, DPLL, iterated conditional modes, simulated annealing and MCMC algorithms that update one variable at a time (Gibbs sampling, single-component Metropolis-Hastings and simulated tempering, etc.), MC-SAT, belief propagation, algorithms for maximum expected utility, etc.

Our approach depends on the concept of "default" values that occur much more frequently than others. In relational domains, the default is false for atoms and true for clauses. In a domain where most variables assume the default value, it is wasteful to allocate memory for all variables and functions in advance. The basic idea is to allocate memory only for a small subset of "active" variables and functions, and activate more if necessary as inference proceeds. In addition to saving memory, this can reduce inference time as well, for we do not allocate memory and compute values for functions that are never used.

Definition 3.1.   Let X be the set of variables and D be their domain.[2] The *default value* $d^* \in D$ is the most frequent value of the variables. An *evidence variable* is a variable whose value is given and fixed. A *function* $f = f(z_1, z_2, \cdots, z_k)$ inputs z_i's, which are either variables or functions, and outputs some value in the range of f.

Although these methods can be applied to other inference algorithms, we focus on relational domains. Variables are ground atoms, which take binary values (i.e., $D = \{true, false\}$). The default value for variables is false (i.e., $d^* = false$). Examples of functions are clauses and DeltaCost in MaxWalkSAT. Like variables, functions may also have default values (e.g., true for clauses). The inputs to a relational inference algorithm are a weighted KB and a set of evidence atoms (DB). Eager algorithms work by first carrying out propositionalization and then calling a propositional algorithm. In lazy inference, we directly work on the KB and DB. The following concepts are crucial to lazy inference.

Definition 3.2.   A variable v is *active* iff v is set to a non-default value at some point, and x is *inactive* if the value of x has always been d^*. A function f is *activated* by a variable v if either v is an input of f, or v activates a function g that is an input of f.

BASIC LAZY INFERENCE

Let $\mathcal{A}$ be the eager algorithm that we want to make lazy. We make three assumptions about $\mathcal{A}$:

1. $\mathcal{A}$ updates one variable at a time. (If not, the extension is straightforward.)

2. The values of variables in $\mathcal{A}$ are properly encapsulated so that they can be accessed by the rest of the algorithm only via two methods: ReadVar(x) (which returns the value of x) and WriteVar(x, v) (which sets the value of x to v). This is reasonable given the conventions in software development, and if not, it is easy to implement.

[2]For simplicity, we assume that all variables have the same domain. The extension to different domains is straightforward.

3. $\mathcal{A}$ always sets values of variables before calling a function that depends on those variables, as it should be.

To develop the lazy version of $\mathcal{A}$, we first identify the variables (usually all) and functions to make lazy. We then modify the value-accessing methods and replace the propositionalization step with lazy initialization as follows. The rest of the algorithm remains the same.

ReadVar(x): If x is in memory, Lazy-$\mathcal{A}$ returns its value as $\mathcal{A}$; otherwise, it returns d^*.

WriteVar(x, v): If x is in memory, Lazy-$\mathcal{A}$ updates its value as $\mathcal{A}$. If not, and if $v = d^*$, no action is taken; otherwise, Lazy-$\mathcal{A}$ activates (allocates memory for) x and the functions activated by x, and then sets the value.

Initialization: Lazy-$\mathcal{A}$ starts by allocating memory for the lazy functions that output non-default values when all variables assume the default values. It then calls WriteVar to set values for evidence variables, which activates those evidence variables with non-default values and the functions they activate. Such variables become the initial active variables and their values are fixed throughout the inference.

Lazy-$\mathcal{A}$ carries out the same inference steps as $\mathcal{A}$ and produces the same result. It never allocates memory for more variables/functions than $\mathcal{A}$, but each access incurs slightly more overhead (in checking whether a variable or function is in memory). In the worst case, most variables are updated, and Lazy-$\mathcal{A}$ produces little savings. However, if the updates are sparse, as is the case for most algorithms in relational domains, Lazy-$\mathcal{A}$ can greatly reduce memory and time because it activates and computes the values for many fewer variables and functions.

REFINEMENTS

The basic version of lazy inference above is generally applicable, but it does not exploit the characteristics of the algorithm and may yield little savings in some cases. Next, we describe three refinements that apply to many algorithms. See Poon *et al.* [108] for additional optimizations and implementation details.

Function activation: To find the functions activated by a variable v, the basic version includes all functions that depend on v by traversing the dependency graph of variables/functions, starting from v. This can be done very efficiently, but may include functions whose values remain the default even if v changes its value. A more intelligent approach traverses the graph and only returns functions whose values become non-default if v changes its value. In general, this requires evaluating the functions, which can be expensive, but for some functions (e.g., clauses), it can be done efficiently as described at the end of this section. Which approach to use depends on the function.

Variable recycling: In the basic version, a variable is activated once it is set to a non-default value, and so are the functions activated by the variable. This is necessary if the variable keeps the non-default value, but could be wasteful in algorithms where many such updates are temporary (e.g., in simulated annealing and MCMC, we may temporarily set a variable to true to compute the probability for flipping). To save memory, if an inactive variable considered for flipping is not flipped in the end, we can discard it along with the functions activated by it. However, this will increase inference time if the variable is considered again later, so we should only apply this if we want to trade off time for memory, or if the variable is unlikely to be considered again.

Smart randomization: Many local-search and MCMC algorithms use randomization to set initial values for variables, or to choose the next variable to update. This is usually done without taking the sparsity of relational domains into account. For example, WalkSAT assigns random initial values to all variables; in a sparse domain where most atoms should be false, this is wasteful, because we will need to flip many atoms back to false. In lazy inference, we can randomize in a better way by focusing more on the active atoms and those "nearby," since they are more likely to become true. Formally, a variable u is a *1-neighbor* of a variable v if u is an input of a function f activated by v; u is a *k-neighbor* of v ($k > 1$) if u is a $(k - 1)$-neighbor of v, or if u is a 1-neighbor of a $(k - 1)$-neighbor of v. For initialization, we only randomize the k-neighbors of initial active variables; for variable selection, we also favor such variables.[3] k can be used to trade off efficiency and solution quality. With large k, this reduces to eager inference. The smaller k is, the more time and memory we can save from reducing the number of activations. However, we also run the risk of missing some relevant variables. If the domain is sparse, as typical relational domains are, the risk is negligible, and empirically we have observed that $k = 1$ greatly improves efficiency without sacrificing solution quality.

LAZYSAT

Applying this method to MaxWalkSAT is fairly straightforward: each ground atom is a variable and each ground clause is a function to be made lazy. Following Singla and Domingos [135], we refer to the resulting algorithm as LazySAT. LazySAT initializes by activating true evidence atoms and initial unsatisfied clauses (i.e., clauses which are unsatisfied when the true evidence atoms are set to true and all other atoms are set to false).[4] At each step in the search, the atom that is flipped is activated, as are any clauses that by definition should become active as a result. While computing DeltaCost(v), if v is active, the relevant clauses are already in memory; otherwise, they will be activated when v is set to true (a necessary step before computing the cost change when v is set to true). Table 3.4 gives pseudocode for LazySAT.

See Figure 3.3 for a sample application of LazySAT based on the Friends and Smokers example. The figure on the left shows the initial state with the evidence Sm(A), Fr(A, B), and Fr(B, C).

---

[3]This does not preclude activation of remote relevant variables, which can happen after their clauses are active, or by random selection (e.g., in simulated annealing).

[4] This differs from MaxWalkSAT, which assigns random values to all atoms. However, the LazySAT initialization is a valid MaxWalkSAT initialization, and the two give very similar results empirically. Given the same initialization, the two algorithms will produce exactly the same results.

Table 3.4: Lazy variant of the MaxWalkSAT algorithm.

---

function LazySAT(KB, DB, m_t, m_f, $target$, p)
  inputs: KB, a weighted knowledge base
        DB, database containing evidence
        m_t, the maximum number of tries
        m_f, the maximum number of flips
        $target$, target solution cost
        p, probability of taking a random step
  output: $soln$, best variable assignment found

for $i \leftarrow 1$ to m_t
  $active\_atoms \leftarrow$ atoms in clauses not satisfied by DB
  $active\_clauses \leftarrow$ clauses activated by $active\_atoms$
  $soln \leftarrow$ a random truth assignment to $active\_atoms$
  $cost \leftarrow$ sum of weights of unsatisfied clauses in $soln$
  for $i \leftarrow 1$ to m_f
    if $cost \leq target$
      return "Success, solution is", $soln$
    $c \leftarrow$ a randomly chosen unsatisfied clause
    if Uniform(0,1) $< p$
      $v_f \leftarrow$ a randomly chosen variable from c
    else
      for each variable v in c
        compute DeltaCost(v), using KB if $v \notin active\_atoms$
      $v_f \leftarrow v$ with lowest DeltaCost(v)
    if $v_f \notin active\_atoms$
      add v_f to $active\_atoms$
      add clauses activated by v_f to $active\_clauses$
    $soln \leftarrow soln$ with v_f flipped
    $cost \leftarrow cost +$ DeltaCost(v_f)
  return "Failure, best assignment is", best $soln$ found

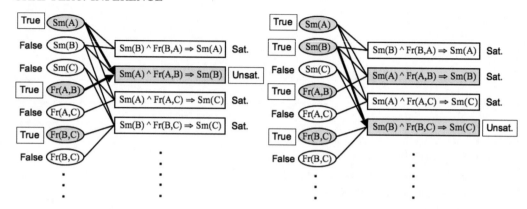

Figure 3.3: Two steps of the LazySAT algorithm. Darkened nodes represent active variables or functions.

By checking all clauses that reference these atoms, we can find and activate the one unsatisfied clause, $Sm(A) \wedge Fr(A, B) \Rightarrow Sm(B)$. LazySAT next selects an unsatisfied clause at random (there is only one so far) and picks an atom to satisfy it using a random or greedy criterion. In this case, LazySAT picks $Sm(B)$ and activates it and the now unsatisfied clause $Sm(B) \wedge Fr(B, C) \Rightarrow Sm(C)$. By again selecting the unsatisfied clause and flipping $Sm(C)$ to satisfy it, LazySAT can arrive at a variable assignment that satisfies all clauses.

Experiments in a number of domains show that LazySAT can yield very large memory reductions, and these reductions increase with domain size [135]. For domains whose full instantiations fit in memory, running time is comparable; as problems become larger, full instantiation for MaxWalk-SAT becomes impossible.

LAZY-MC-SAT
Recall the MC-SAT algorithm from Section 3.2, with pseudocode in Table 3.3. MC-SAT initializes with a solution to all hard clauses found by WalkSAT. In each iteration, it generates a set of clauses M by sampling from the currently satisfied clauses according to their weights. It then calls SampleSAT [150] to select the next state by sampling near-uniformly the solutions to M. SampleSAT initializes with a random state and, in each iteration, performs a simulated annealing step with probability p and a WalkSAT step with probability $1 - p$. To improve efficiency, MC-SAT runs unit propagation in M before calling SampleSAT (i.e., it repeatedly sets atoms in unit clauses to the corresponding fixed values and simplifies the remaining clauses).

Here, the functions to be made lazy are clauses, clauses' memberships in M, and whether an atom is fixed by unit propagation. Effectively, Lazy-MC-SAT initializes by calling LazySAT to find a solution to all hard clauses. At each iteration, it computes M-memberships for and runs unit propagation among active clauses only. It then calls Lazy-SampleSAT, the lazy version of

SampleSAT. Lazy-SampleSAT initializes using smart randomization with $k = 1$. When an inactive atom v is set to false, no action is taken. When it is set to true, Lazy-SampleSAT activates v and the clauses activated by v; it then computes M-memberships for these clauses,[5] and runs unit propagation among them; if v is fixed to false by unit propagation, Lazy-SampleSAT sets v back to false and applies variable recycling. It also applies variable recycling to simulated annealing steps. Notice that, when an M-membership or fixed-atom flag is set, this is remembered until the current run of Lazy-SampleSAT ends, to avoid inconsistencies.

3.4 LIFTED INFERENCE

The inference methods discussed so far are purely probabilistic in the sense that they propositionalize all atoms and clauses and apply standard probabilistic inference algorithms. A key property of first-order logic is that it allows *lifted* inference, where queries are answered without materializing all the objects in the domain (e.g., resolution [122]). Lifted inference is potentially much more efficient than propositionalized inference, and extending it to probabilistic logical languages is a desirable goal. The first approach to lifted probabilistic inference, first-order variable elimination (FOVE), was developed by Poole [104] and extended by Braz *et al.* [11, 12] and Milch *et al.* [85]. (Limited lifted aspects are present in some earlier systems, like Pfeffer *et al.*'s [101] SPOOK.) Unfortunately, variable elimination has exponential cost in the treewidth of the graph, making it infeasible for most real-world applications. Scalable approximate algorithms for probabilistic inference fall into three main classes: loopy belief propagation (BP), Monte Carlo methods, and variational methods.

We have developed a lifted version of BP [138], building on the work of Jaimovich *et al.* [51]. Jaimovich *et al.* pointed out that, if there is no evidence, BP in probabilistic logical models can be trivially lifted, because all groundings of the same atoms and clauses become indistinguishable. Our approach proceeds by identifying the subsets of atoms and clauses that remain indistinguishable even after evidence is taken into account. We then form a network with *supernodes* and *superfeatures* corresponding to these sets, and apply BP to it. This network can be vastly smaller than the full ground network, with the corresponding efficiency gains. Our algorithm produces the unique minimal lifted network for every inference problem.

BELIEF PROPAGATION

In our description of Markov networks from Chapter 2, the graph G contains a node for each variable and an edge connecting each pair of variables that share a potential function, or *factor*. Markov networks can also be represented as *factor graphs* [62]. A factor graph is a bipartite graph with a node for each variable and factor in the model. (For convenience, we will consider one factor $f_i(\mathbf{x}) = \exp(w_i g_i(\mathbf{x}))$ per feature $g_i(\mathbf{x})$, i.e., we will not aggregate features over the same variables into a single factor.) Variables and the factors they appear in are connected by undirected edges. See Figure 3.4 for an example of a factor graph.

---

[5]These clauses must be previously satisfied; otherwise, they would have been activated before.

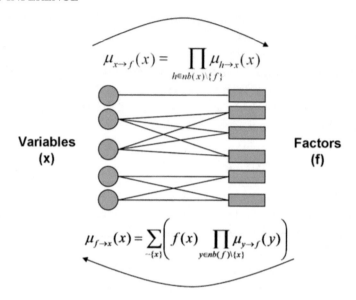

Figure 3.4: Belief propagation in a factor graph. Above is the equation for messages from variables to factors; below is the equation for messages from factors to variables.

The main inference task in graphical models is to compute the conditional probability of some variables (the query) given the values of some others (the evidence), by summing out the remaining variables. This problem is #P-complete, but becomes tractable if the graph is a tree. In this case, the marginal probabilities of the query variables can be computed in polynomial time by *belief propagation*, which consists of passing messages from variable nodes to the corresponding factor nodes and vice-versa. The message from a variable x to a factor f is

$$\mu_{x \to f}(x) = \prod_{h \in nb(x)\setminus\{f\}} \mu_{h \to x}(x) \qquad (3.4)$$

where $nb(x)$ is the set of factors x appears in. The message from a factor to a variable is

$$\mu_{f \to x}(x) = \sum_{\sim\{x\}} \left(f(\mathbf{x}) \prod_{y \in nb(f)\setminus\{x\}} \mu_{y \to f}(y) \right) \qquad (3.5)$$

where $nb(f)$ are the arguments of f, and the sum is over all of these except x. The messages from leaf variables are initialized to 1, and a pass from the leaves to the root and back to the leaves suffices. The (unnormalized) marginal of each variable x is then given by $\prod_{h \in nb(x)} \mu_{h \to x}(x)$. Evidence is incorporated by setting $f(\mathbf{x}) = 0$ for states $\mathbf{x}$ that are incompatible with it. This algorithm can still be applied when the graph has loops (e.g., Figure 3.4), repeating the message-passing until convergence.

Although this *loopy* belief propagation has no guarantees of convergence or of giving accurate results, in practice it often does, and can be much more efficient than other methods. Different schedules may be used for message-passing. Here we assume *flooding*, the most widely used and generally best-performing method, in which messages are passed from each variable to each corresponding factor and back at each step (after initializing all variable messages to 1).

Belief propagation can also be used for exact inference in arbitrary graphs, by combining nodes until a tree is obtained, but this suffers from the same combinatorial explosion as variable elimination.

LIFTED BELIEF PROPAGATION

We begin with some necessary definitions. These assume the existence of an MLN L, set of constants C, and evidence database E (set of ground literals). For simplicity, our definitions and explanation of the algorithm will assume that each predicate appears at most once in any given MLN clause. We will then describe how to handle multiple occurrences of a predicate in a clause.

Definition 3.3.   A *supernode* is a set of groundings of a predicate that all send and receive the same messages at each step of belief propagation, given L, C and E. The supernodes of a predicate form a partition of its groundings.

A *superfeature* is a set of groundings of a clause that all send and receive the same messages at each step of belief propagation, given L, C and E. The superfeatures of a clause form a partition of its groundings.

Definition 3.4.   A *lifted network* is a factor graph composed of supernodes and superfeatures. The factor corresponding to a superfeature $g(x)$ is $\exp(wg(x))$, where w is the weight of the corresponding first-order clause. A supernode and a superfeature have an edge between them iff some ground atom in the supernode appears in some ground clause in the superfeature. Each edge has a positive integer weight. A *minimal lifted network* is a lifted network with the smallest possible number of supernodes and superfeatures.

The first step of lifted BP is to construct the minimal lifted network. The size of this network is $O(nm)$, where n is the number of supernodes and m the number of superfeatures. In the best case, the lifted network has the same size as the MLN L; in the worst case, as the ground Markov network $M_{L,C}$.

The second and final step in lifted BP is to apply standard BP to the lifted network, with two changes:

1. The message from supernode x to superfeature f becomes

$$\mu_{f \to x}^{n(f,x)-1} \prod_{h \in nb(x) \setminus \{f\}} \mu_{h \to x}(x)^{n(h,x)}$$

where $n(h, x)$ is the weight of the edge between h and x.

2. The (unnormalized) marginal of each supernode (and, therefore, of each ground atom in it) is given by $\prod_{h \in nb(x)} \mu_{h \to x}^{n(h,x)}(x)$.

The weight of an edge is the number of identical messages that would be sent from the ground clauses in the superfeature to each ground atom in the supernode if BP was carried out on the ground network. The $n(f, x) - 1$ exponent reflects the fact that a variable's message to a factor excludes the factor's message to the variable.

The lifted network is constructed by (essentially) simulating BP and keeping track of which ground atoms and clauses send the same messages. Initially, the groundings of each predicate fall into three groups: known true, known false and unknown. (One or two of these may be empty.) Each such group constitutes an initial supernode. All groundings of a clause whose atoms have the same combination of truth values (true, false or unknown) now send the same messages to the ground atoms in them. In turn, all ground atoms that receive the same number of messages from the superfeatures they appear in send the same messages, and constitute a new supernode. As the effect of the evidence propagates through the network, finer and finer supernodes and superfeatures are created.

If a clause involves predicates $R_1, \ldots, R_k$, and $N = (N_1, \ldots, N_k)$ is a corresponding tuple of supernodes, the groundings of the clause generated by N are found by joining $N_1, \ldots, N_k$ (i.e., by forming the Cartesian product of the relations $N_1, \ldots, N_k$, and selecting the tuples in which the corresponding arguments agree with each other, and with any corresponding constants in the first-order clause). Conversely, the groundings of predicate R_i connected to elements of a superfeature F are obtained by projecting F onto the arguments it shares with R_i. Lifted network construction thus proceeds by alternating between two steps:

1. Form superfeatures by doing joins of their supernodes.

2. Form supernodes by projecting superfeatures down to their predicates, and merging atoms with the same projection counts.

Pseudocode for the algorithm is shown in Table 3.5. The projection counts at convergence are the weights associated with the corresponding edges.

To handle clauses with multiple occurrences of a predicate, we keep a tuple of edge weights, one for each occurrence of the predicate in the clause. A message is passed for each occurrence of the predicate, with the corresponding edge weight. Similarly, when projecting superfeatures into supernodes, a separate count is maintained for each occurrence, and only tuples with the same counts for all occurrences are merged.

Figure 3.5 demonstrates how the initial supernodes and superfeatures are constructed for a friendship ($Fr(x, y)$) and smoking ($Sm(x)$) domain based on the example of Table 2.1. The only rule in the MLN is $Sm(x) \land Fr(x, y) \Rightarrow Sm(y)$ (if two people are friends and one smokes, so does the other). We assume a set of n constants, $\{A, B, C, D, \ldots\}$ and the evidence $Sm(A)$, $Fr(B, C)$, and $Fr(C, B)$. To form the supernodes, each predicate is partitioned into its unknown and true groundings (since there are no false atoms in the evidence), yielding four initial supernodes. These are joined in

Table 3.5: Lifted network construction algorithm.

---

function LNC(L, C, E)
   inputs: L, a Markov logic network
          C, a set of constants
          E, a set of ground literals
   output: M, a lifted network
for each predicate P
   for each truth value t in {*true, false, unknown*}
     form a supernode containing all groundings of P with truth value t
repeat
   for each clause involving predicates $P_1, \ldots, P_k$
     for each tuple of supernodes $(N_1, \ldots, N_k)$,
       where N_i is a P_i supernode
       form a superfeature F by joining $N_1, \ldots, N_k$
   for each predicate P
     for each superfeature F it appears in
       $S(P, F) \leftarrow$ projection of the tuples in F down to the variables in P
       for each tuple s in $S(P, F)$
         $T(s, F) \leftarrow$ number of F's tuples that were projected into s
       $S(P) \leftarrow \bigcup_F S(P, F)$
       form a new supernode from each set of tuples in $S(P)$ with the
        same $T(s, F)$ counts for all F
until convergence
add all current supernodes and superfeatures to M
for each supernode N and superfeature F in M
   add to M an edge between N and F with weight $T(s, F)$
return M

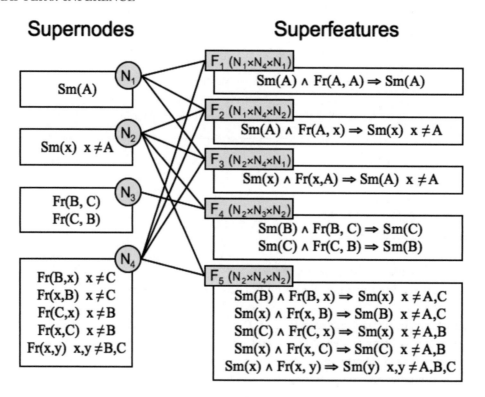

Figure 3.5: Initial supernodes and superfeatures when Sm(A), Fr(B, C), and Fr(C, B) are evidence and the only formula is Sm(x) ∧ Fr(x, y) ⇒ Sm(y).

all possible ways to produce five superfeatures. For clarity, each superfeature in Figure 3.5 is labeled with the supernodes it joins. The next step in lifted network construction is to count the number of times each atom appears in each position in each superfeature, as shown in Table 3.6 for the Sm() predicate. Atoms with the same counts are clustered into new supernodes, as indicated by the horizontal lines in Table 3.6. These new supernodes are joined to produce new superfeatures, then projected to produce refined supernodes, and so on until convergence.

Theorem 3.5.   *Given an MLN L, set of constants C and set of ground literals E, there exists a unique minimal lifted network M*, and algorithm LNC(L, C, E) returns it. Belief propagation applied to M* produces the same results as belief propagation applied to the ground Markov network generated by L and C.*

See Singla and Domingos [138] for the proof.

Clauses involving evidence atoms can be simplified (false literals and clauses containing true literals can be deleted). As a result, duplicate clauses may appear, and the corresponding superfeatures

Table 3.6: Projection of superfeatures from Figure 3.5 down to the Sm() predicate. Two projected counts are provided for each feature, since Sm() appears twice in the rule $Sm(x) \wedge Fr(x, y) \Rightarrow Sm(y)$. Horizontal lines cluster the atoms into a new set of supernodes.

Atom	F_1		F_2		F_3		F_4		F_5	
Sm(A)	1	1	$n-1$	0	0	$n-1$	0	0	0	0
Sm(B)	0	0	0	1	1	0	1	1	$n-2$	$n-2$
Sm(C)	0	0	0	1	1	0	1	1	$n-2$	$n-2$
Sm(D)	0	0	0	1	1	0	0	0	$n-1$	$n-1$
Sm(E)	0	0	0	1	1	0	0	0	$n-1$	$n-1$
...	0	0	0	1	1	0	0	0	$n-1$	$n-1$

can be merged. This will typically result in duplicate instances of tuples. Each tuple in the merged superfeature is assigned a weight $\sum_i m_i w_i$, where m_i is the number of duplicate tuples resulting from the ith superfeature and w_i is the corresponding weight. During the creation of supernodes, $T(s, F)$ is now the number of F tuples projecting into s multiplied by the corresponding weight. This can greatly reduce the size of the lifted network. When no evidence is present, the algorithm reduces to the one proposed by Jaimovich *et al.* [51].

An important question remains: how to represent supernodes and superfeatures. Although this does not affect the space or time cost of belief propagation (where each supernode and superfeature is represented by a single symbol), it can greatly affect the cost of constructing the lifted network. The simplest option is to represent each supernode or superfeature *extensionally* as a set of tuples (i.e., a relation), in which case joins and projections reduce to standard database operations. However, in this case the cost of constructing the lifted network is similar to the cost of constructing the full ground network, and can easily become the bottleneck. A better option is to use a more compact *intensional* representation, as done by Poole [104] and Braz *et al.* [11, 12].[6]

A ground atom can be viewed as a first-order atom with all variables constrained to be equal to constants, and similarly for ground clauses. (For example, R(A, B) is R(x, y) with x = A and y = B.) We represent supernodes by sets of (α, γ) pairs, where α is a first-order atom and γ is a set of constraints, and similarly for superfeatures. Constraints are of the form $x = y$ or $x \neq y$, where x is an argument of the atom and y is either a constant or another argument. For example, (S(v, w, x, y, z), {w = x, y = A, z ≠ B, z ≠ C}) compactly represents all groundings of S(v, w, x, y, z) compatible with the constraints. Notice that variables may be left unconstrained, and that infinite sets of atoms can be finitely represented in this way.

Let the *default value* of a predicate R be its most frequent value given the evidence (true, false or unknown). Let $S_{R,i}$ be the set of constants that appear as the ith argument of R only in

---

[6]Superfeatures are related, but not identical, to the parfactors of Poole and Braz *et al.*. One important difference is that superfeatures correspond to factors in the original graph, while parfactors correspond to factors created during variable elimination. Superfeatures are thus exponentially more compact.

groundings with the default value. Supernodes not involving any members of $S_{R,i}$ for any argument i are represented extensionally (i.e. with pairs (α, γ) where γ contains a constraint of the form $\text{x} = \text{A}$, where A is a constant, for each argument x). Initially, supernodes involving members of $\mathbf{S}_{R,i}$ are represented using (α, γ) pairs containing constraints of the form $\text{x} \neq \text{A}$ for each $\text{A} \in C \setminus S_{R,i}$.[7] When two or more supernodes are joined to form a superfeature F, if the kth argument of F's clause is the $i(j)$th argument of its jth literal, $S_k = \bigcap_j S_{r(j),i}$, where $r(j)$ is the predicate symbol in the jth literal. F is now represented analogously to the supernodes, according to whether or not it involves elements of $\mathbf{S}_k$. If F is represented intensionally, each (α, γ) pair is divided into one pair for each possible combination of equality/inequality constraints among the clause's arguments, which are added to γ. When forming a supernode from superfeatures, the constraints in each (α, γ) pair in the supernode are the union of (a) the corresponding constraints in the superfeatures on the variables included in the supernode, and (b) the constraints induced by the excluded variables on the included ones. This process is analogous to the shattering process of Braz *et al.* [11].

In general, finding the most compact representation for supernodes and superfeatures is an intractable problem, and is an area of active research.

FURTHER READING

The MaxWalkSAT algorithm used for MPE inference in Markov logic is described in Kautz *et al.* [53]. Its source code is available at http://www.cs.rochester.edu/~kautz/walksat/. MaxWalkSAT is a propositional MaxSAT solver, so the MLN formulas are first grounded in all possible ways to construct the MaxSAT problem for MaxWalkSAT to solve.

For full details on MC-SAT, including experimental evaluation, see Poon and Domingos [105].

For more information about LazySAT, including experiments, see Singla and Domingos [135]. For a more general treatment of lazy inference, including details on Lazy-MC-SAT and how to make other algorithms lazy, see Poon *et al.* [108].

Recently, Riedel [118] proposed Cutting Plane Inference (CPI), an alternative MAP solver based on cutting plane methods. CPI is similar to lazy inference in that it adds clauses only as needed. Riedel uses both MaxWalkSAT and integer linear programming as the "base" solver. On a semantic role labeling problem, CPI with integer linear programming was able to find the exact MAP solution faster than MaxWalkSAT could approximate it.

An introduction to the original belief propagation algorithm can be found in Yedidia *et al.* [156]. For details, proofs, and experiments on lifted belief propagation, see Singla and Domingos [138]. Publications on exact lifted inference include Poole [104], Braz *et al.* [11, 12], and Milch *et al.* [85].

---

[7]In practice, variables are typed, and C is replaced by the domain of the argument; and the set of constraints is only stored once, and pointed to as needed.

CHAPTER 4

Learning

Learning algorithms use data to automatically construct or refine a Markov logic network, typically obtaining much better models with less work than manually specifying the complete model. We begin by discussing weight learning, in which we try to find the formula weights that maximize the likelihood or conditional likelihood of a relational database. Our methods for solving this problem are based on convex optimization but take into account the hardness of inference. Structure learning, in which we try to learn the formulas themselves, is a much harder problem. We present two approaches based on inductive logic programming. Finally, we describe how new concepts can be discovered through statistical predicate invention, and how experience in one domain can help us learn faster in another through transfer learning. Solutions for these problems naturally use *second-order Markov logic* to reason about predicates in much the way that first-order Markov logic reasons about objects.

4.1 WEIGHT LEARNING

GENERATIVE WEIGHT LEARNING

MLN weights can be learned generatively by maximizing the likelihood of a relational database (Equation 2.3). This relational database consists of one or more "possible worlds" that form our training examples. We assume that the set of constants of each type is known. We also make a closed-world assumption [37]: all ground atoms not in the database are false.[1] If there are n possible ground atoms, a database is effectively a vector $x = (x_1, \ldots, x_l, \ldots, x_n)$ where x_l is the truth value of the lth ground atom ($x_l = 1$ if the atom appears in the database, and $x_l = 0$ otherwise). Note that we can learn to generalize from even a single example because the clause weights are shared across their many respective groundings. Given a database, MLN weights can in principle be learned using standard methods, as follows. If the ith formula has $n_i(x)$ true groundings in the data x, then by Equation 2.3, the derivative of the log-likelihood with respect to its weight is

$$\frac{\partial}{\partial w_i} \log P_w(X=x) = n_i(x) - \sum_{x'} P_w(X=x') \, n_i(x') \tag{4.1}$$

where the sum is over all possible databases x', and $P_w(X=x')$ is $P(X=x')$ computed using the current weight vector $w = (w_1, \ldots, w_i, \ldots)$. In other words, the ith component of the gradient is simply the difference between the number of true groundings of the ith formula in the data and its expectation according to the current model. We can derive this expression as follows. The

---

[1]This assumption can be removed by using the expectation maximization (EM) algorithm [25] to learn from the resulting incomplete data, as we will see in Section 4.3.

log-likelihood itself is just the log of Equation 2.3:

$$P_w(X=x) = \frac{1}{Z} \exp\left(\sum_j w_j n_j(x)\right)$$

$$\log P_w(X=x) = \sum_j w_j n_j(x) - \log Z \tag{4.2}$$

Now we take the partial derivative.

$$\frac{\partial}{\partial w_i} \log P_w(X=x) = \frac{\partial}{\partial w_i} \sum_j w_j n_j(x) - \frac{\partial}{\partial w_i} \log Z$$

$$= n_i(x) - \frac{1}{Z} \frac{\partial}{\partial w_i} Z \tag{4.3}$$

$$= n_i(x) - \frac{1}{Z} \sum_{x'} \frac{\partial}{\partial w_i} \exp\left(\sum_j w_j n_j(x')\right) \tag{4.4}$$

$$= n_i(x) - \frac{1}{Z} \sum_{x'} \exp\left(\sum_j w_j n_j(x')\right) n_i(x') \tag{4.5}$$

$$= n_i(x) - \sum_{x'} P_w(X=x') n_i(x')$$

Equation 4.3 follows from basic derivative rules. In particular, note that the derivative of $w_j n_j(x)$ with respect to w_i is zero for $j \neq i$. We substitute the definition of Z to obtain Equation 4.4, and we apply the chain and sum rules to obtain Equation 4.5. By moving Z inside the summation and substituting the definition of P_w, we arrive at Equation 4.1.

In the worst case, counting the number of true groundings of a formula in a database is intractable, even when the formula is a single clause, as stated in the following proposition (due to Dan Suciu):

Proposition 4.1. *Counting the number of true groundings of a first-order clause in a database is #P-complete in the length of the clause.*

Proof. Counting satisfying assignments of propositional monotone 2-CNF is #P-complete [123]. This problem can be reduced to counting the number of true groundings of a first-order clause in a database as follows. Consider a database composed of the ground atoms $R(0, 1)$, $R(1, 0)$ and $R(1, 1)$. Given a monotone 2-CNF formula, construct a formula F that is a conjunction of atoms of the form $R(x_i, x_j)$, one for each disjunction $x_i \vee x_j$ appearing in the CNF formula. (For example, $(x_1 \vee x_2) \wedge (x_3 \vee x_4)$ would yield $R(x_1, x_2) \wedge R(x_3, x_4)$.) There is a one-to-one correspondence between the satisfying assignments of the 2-CNF and the true groundings of F. The latter are the

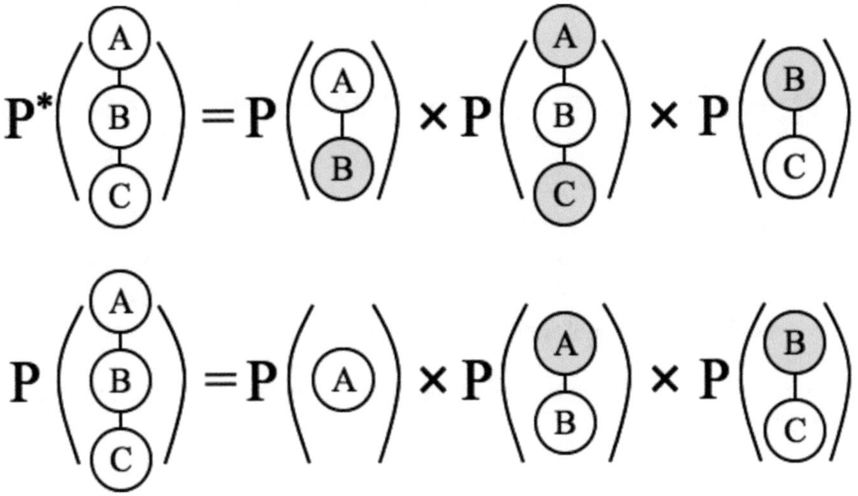

Figure 4.1: Comparison of pseudo-likelihood and likelihood on a simple three-node network. Shaded nodes represent conditioning on the true values. Pseudo-likelihood (above) is the product of the probability of each node conditioned on its Markov blanket. In contrast, likelihood (below) obeys the chain rule of probability when decomposed.

false groundings of the disjunction of the negations of all the $R(x_i, x_j)$, and thus can be counted by counting the number of true groundings of this clause and subtracting it from the total number of groundings. □

In large domains, the number of true groundings of a formula may be counted approximately, by uniformly sampling groundings of the formula and checking whether they are true in the data. In smaller domains, and in most applications to date, we use an efficient recursive algorithm to find the exact count.

A second problem with Equation 4.1 is that computing the expected number of true groundings is also intractable, requiring inference over the model. Instead, we maximize the pseudo-likelihood of the data, a widely used alternative [6]. If x is a possible world (relational database) and x_l is the lth ground atom's truth value, the pseudo-log-likelihood of x given weights w is

$$\log P_w^*(X = x) = \sum_{l=1}^{n} \log P_w(X_l = x_l | MB_x(X_l)) \qquad (4.6)$$

where $MB_x(X_l)$ is the state of X_l's Markov blanket in the data (i.e., the truth values of the ground atoms it appears in some ground formula with). Figure 4.1 illustrates the difference between pseudo-likelihood and likelihood on a simple three-node network. The gradient of the pseudo-log-likelihood is

$$\frac{\partial}{\partial w_i} \log P_w^*(X=x) = \sum_{l=1}^{n} [n_i(x) - P_w(X_l=0|MB_x(X_l)) \, n_i(x_{[X_l=0]})$$
$$- P_w(X_l=1|MB_x(X_l)) \, n_i(x_{[X_l=1]})] \qquad (4.7)$$

where $n_i(x_{[X_l=0]})$ is the number of true groundings of the ith formula when we force $X_l = 0$ and leave the remaining data unchanged, and similarly for $n_i(x_{[X_l=1]})$. Computing this expression (or Equation 4.6) does not require inference over the model, and is therefore much faster. The computation can be made more efficient in several ways:

- The sum in Equation 4.7 can be greatly sped up by ignoring predicates that do not appear in the ith formula.

- The counts $n_i(x)$, $n_i(x_{[X_l=0]})$ and $n_i(x_{[X_l=0]})$ do not change with the weights, and need only be computed once (as opposed to in every iteration of optimization).

- Ground formulas whose truth value is unaffected by changing the truth value of any single literal may be ignored, since then $n_i(x) = n_i(x_{[X_l=0]}) = n_i(x_{[X_l=0]})$. In particular, this holds for any clause which contains at least two true literals. This can often be the great majority of ground clauses.

Combined with the L-BFGS optimizer [71], pseudo-likelihood yields efficient learning of MLN weights even in domains with millions of ground atoms [117]. However, the pseudo-likelihood parameters may lead to poor results when long chains of inference are required. In the next section on discriminative learning, we will see how optimizing the conditional likelihood instead can yield better results.

In order to reduce overfitting, we penalize each weight with a Gaussian prior. We apply this strategy not only to generative learning, but to all of our weight learning methods, including those embedded within structure learning.

DISCRIMINATIVE WEIGHT LEARNING

In discriminative learning, we know *a priori* which atoms will be evidence and which ones will be queried, and the goal is to correctly predict the latter given the former. Such a distinction between evidence and query naturally exists in most applications. For example, in collective classification (Section 6.1), the goal is to predict each object's class given all other attributes and relations. If we partition the ground atoms in the domain into a set of evidence atoms X and a set of query atoms Y, the *conditional likelihood* of Y given X is

$$P(Y=y|X=x) = \frac{1}{Z_x} \exp\left(\sum_i w_i n_i(x, y) \right) \qquad (4.8)$$

where Z_x normalizes over all possible worlds consistent with the evidence x, and $n_i(x, y)$ is the number of true groundings of the ith formula in the data. Note that ground clauses satisfied by the evidence or having no query atoms can be ignored, since no configuration of the query atoms can change the truth value of these clauses.

Weights are learned discriminatively by maximizing the conditional log-likelihood (CLL). CLL is easier to optimize than log-likelihood because the evidence typically constrains the probability distribution to a much smaller set of likely states. Singla and Domingos [133] demonstrated that learning weights to optimize the CLL can greatly outperform pseudo-likelihood learning for MLNs. However, note that CLL and pseudo-likelihood are equivalent for the special case where the Markov blanket of each query atom is in the evidence, such as in the logistic regression example from Chapter 2.

Since most of the optimization literature assumes that the objective function is to be minimized, we will equivalently minimize the negative conditional log likelihood. We discuss two classes of optimization algorithms, first-order and second-order methods. First-order methods pick a search direction based on the function's gradient. Second-order methods derive a search direction from approximating the function as a quadratic surface. While this approximation may not hold globally, it can be good enough in local regions to greatly speed up convergence.

Voted Perceptron

Gradient descent algorithms use the gradient, $\mathbf{g}$, scaled by a learning rate, η, to update the weight vector $\mathbf{w}$ in each step:

$$\mathbf{w}_{t+1} = \mathbf{w}_t - \eta \mathbf{g} \tag{4.9}$$

See Figure 4.2 for a simple illustration. Note that the gradient is always perpendicular to the level sets.

In an MLN, the derivative of the negative CLL with respect to a weight is the difference of the expected number of true groundings of the corresponding clause and the actual number according to the data:

$$\frac{\partial}{\partial w_i}(-\log P_w(Y=y|X=x)) = -n_i(x, y) + \sum_{y'} P_w(Y=y'|X=x)\, n_i(x, y') \tag{4.10}$$

$$= E_{w,y}[n_i(x, y)] - n_i(x, y) \tag{4.11}$$

where y is the state of the non-evidence atoms in the data, x is the state of the evidence, and the expectation $E_{w,y}$ is over the non-evidence atoms Y. We use the w, y subscript to emphasize that the expectation depends on the current weight vector, w, and is obtained by summing over states of the world y. In the remainder of this section, we will write expectations $E_{w,y}[n_i(x, y)]$ as $E_w[n_i]$ when the meaning is clear from the context.

The basic idea of the voted perceptron (VP) algorithm [15] is to approximate the intractable expectations $E_w[n_i]$ with the counts in the MPE (most probable explanation) state, which is the most probable state of non-evidence atoms given the evidence. To combat overfitting, instead of

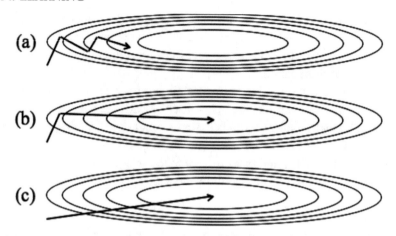

Figure 4.2: Behavior of (a) gradient descent, (b) conjugate gradient, and (c) Newton's method on a quadratic surface. The surface is presented as its level sets.

returning the final weights, VP returns the average of the weights from all iterations of gradient descent.

Collins originally proposed VP for training hidden Markov models discriminatively, and in this case the MPE state is unique and can be computed exactly in polynomial time using the Viterbi algorithm. As discussed in Section 3.1, MPE inference in MLNs is intractable but can be reduced to solving a weighted maximum satisfiability problem, for which efficient algorithms exist, such as MaxWalkSAT [53]. Singla and Domingos [133] use this approach and discuss how the resulting algorithm can be viewed as approximately optimizing log-likelihood. However, the use of voted perceptron in MLNs is potentially complicated by the fact that the MPE state may no longer be unique, and MaxWalkSAT is not guaranteed to find it.

Contrastive Divergence

The contrastive divergence (CD) algorithm is similar to VP, except that it approximates the expectations $E_w[n_i]$ from a small number of MCMC samples instead of using the MPE state. Using MCMC is presumably more accurate and stable, since it converges to the true expectations in the limit. While running an MCMC algorithm to convergence at each iteration of gradient descent is infeasibly slow, Hinton [47] has shown that a few iterations of MCMC yield enough information to choose a good direction for gradient descent. Hinton named this method *contrastive divergence* because it can be interpreted as optimizing a difference of Kullback-Leibler divergences. Contrastive divergence can also be seen as an efficient way to approximately optimize log-likelihood.

The MCMC algorithm typically used with contrastive divergence is Gibbs sampling, but for MLNs the much faster alternative of MC-SAT (Section 3.2) is available. Because successive samples in MC-SAT are much less correlated than successive sweeps in Gibbs sampling, they carry more information and are likely to yield a better descent direction. In particular, the different samples are likely to be from different modes, reducing the error and potential instability associated with choosing a single mode.

In an MLN, each state typically contains a large number of instances of each clause, unlike i.i.d. learning, which would contain only one. This makes even a small number of MCMC steps particularly effective in relational domains. In our experience, five samples is often sufficient and additional samples may not be worth the time: any increased accuracy that 10 or 100 samples might bring is offset by the increased time per iteration. We avoid the need for burn-in by starting at the last state sampled in the previous iteration of gradient descent. (This differs from Hinton's approach, which always starts at the true values in the training data.)

Per-Weight Learning Rates

VP and CD are both simple gradient descent procedures, and as a result highly vulnerable to the problem of ill-conditioning. On ill-conditioned problems, the direction of steepest descent is nearly orthogonal to the direction to the optimum, causing gradient descent to take many iterations to converge. Ill-conditioning occurs when the *condition number*, the ratio between the largest and smallest eigenvalues of the Hessian (matrix of second derivatives), is far from one. In MLNs, the Hessian is the negative covariance matrix of the clause counts. Because some clauses can have vastly greater numbers of true groundings than others, the variances of their counts can be correspondingly larger, and ill-conditioning becomes a serious issue. Intuitively, gradient descent is slow because no single learning rate is appropriate for all weights; a clause with $O(n)$ groundings, such as $\texttt{Smokes(x)} \Rightarrow \texttt{Cancer(x)}$, will typically require a much higher learning rate than a clause with $O(n^3)$ groundings, such as transitivity.

One solution is to modify both algorithms to have a different learning rate for each weight. Since tuning every learning rate separately is impractical, we use a simple heuristic to assign a learning rate to each weight:

$$\eta_i = \frac{\eta}{n_i}$$

where η is the user-specified global learning rate and n_i is the number of true groundings of the ith formula. (To avoid dividing by zero, if $n_i = 0$ then $\eta_i = \eta$.) When computing this number, we ignore the groundings that are satisfied by the evidence (e.g., $A \Rightarrow B$ when A is false). This is because, being fixed, they cannot contribute to the variance.

Diagonal Newton

When the function being optimized is quadratic, as in Figure 4.2, Newton's method can move to the global minimum or maximum in a single step. It does so by multiplying the gradient, $\mathbf{g}$, by the inverse Hessian, $\mathbf{H}^{-1}$:

$$\mathbf{w}_{t+1} = \mathbf{w}_t - \mathbf{H}^{-1}\mathbf{g} \tag{4.12}$$

When there are many weights, inverting the full Hessian becomes infeasible. A common approximation is to use the *diagonal* Newton (DN) method, which makes the simplifying assumption that all off-diagonal entries of the Hessian are zero, yielding an easily inverted diagonal matrix. DN typically uses a smaller step size than the full Newton method. This is important when applying the algorithm to non-quadratic functions, such as the CLL of an MLN, where the quadratic approximation is only good within a local region.

The Hessian of the negative CLL for an MLN is simply the covariance matrix:

$$\frac{\partial}{\partial w_i \partial w_j}(-\log P_w(Y=y|X=x)) = E_w[n_i n_j] - E_w[n_i]E_w[n_j] \tag{4.13}$$

Like the gradient, this can be estimated using samples from MC-SAT. Since the off-diagonal entries capture interactions between different clauses, the diagonal assumption is equivalent to assuming that all clauses are uncorrelated. This can be a reasonable approximation if the true correlations are weak. In each iteration, we take a step in the diagonal Newton direction:

$$w_i = w_i - \alpha \frac{E_w[n_i] - n_i}{E_w[n_i^2] - (E_w[n_i])^2} \tag{4.14}$$

The step size α can be computed in a number of ways, including keeping it fixed, but we have obtained the best results using the following method. Given a search direction $\mathbf{d}$ and Hessian matrix $\mathbf{H}$, we compute the step size as follows:

$$\alpha = \frac{-\mathbf{d}^T \mathbf{g}}{\mathbf{d}^T \mathbf{H} \mathbf{d} + \lambda \mathbf{d}^T \mathbf{d}} \tag{4.15}$$

where $\mathbf{d}$ is the search direction. For a quadratic function and $\lambda = 0$, this step size would move to the minimum function value along $\mathbf{d}$. Since our function is not quadratic, a non-zero λ term serves to limit the size of the step to a region in which our quadratic approximation is good. After each step, we adjust λ to increase or decrease the size of the so-called *model trust region* [95] based on how well the approximation matched the function. Let Δ_{actual} be the actual change in the function value, and let Δ_{pred} be the predicted change in the function value from the previous gradient and Hessian and our last step, $\mathbf{d}_{t-1}$:

$$\Delta_{pred} = \mathbf{d}_{t-1}^T \mathbf{g}_{t-1} + 1/2 \, \mathbf{d}_{t-1}^T \mathbf{H}_{t-1} \mathbf{d}_{t-1} \tag{4.16}$$

A standard method for adjusting λ is as follows [35]:

$$if(\Delta_{actual}/\Delta_{pred} > 0.75) \quad then \quad \lambda_{t+1} = \lambda_t/2$$
$$if(\Delta_{actual}/\Delta_{pred} < 0.25) \quad then \quad \lambda_{t+1} = 4\lambda_t$$

Since we cannot efficiently compute the actual change in negative CLL, we approximate it as the product of the step we just took and the gradient after taking it: $\Delta_{actual} = \mathbf{d}_{t-1}^T \mathbf{g}_t$. Since the negative CLL is a convex function, this product is an upper bound on the actual change. This bound is often quite loose. Suppose the last step took us directly to the optimum. At the optimum, the new gradient g_t is zero, so Δ_{actual} is approximated as zero regardless of the actual change in function value. Therefore, we only increase λ when Δ_{actual} is positive. In this case, the CLL may be worse than before, so the step is rejected and redone after adjusting λ.

In models with thousands of weights or more, storing the entire Hessian matrix becomes impractical. However, when the Hessian appears only inside a quadratic form, as above, we can avoid storing the matrix by computing the values of these forms directly from the clause counts in each sampled state:

$$\mathbf{d}^T \mathbf{H} \mathbf{d} = E_w[(\textstyle\sum_i d_i n_i)^2] - (E_w[\textstyle\sum_i d_i n_i])^2 \tag{4.17}$$

As with the Equations 4.11 and 4.13, these expectations are over all possible worlds but can be approximated by a small number of samples. The product of the Hessian by a vector can also be computed compactly [100]. Note that α is computed using the full Hessian matrix, but the step direction is computed from the diagonal approximation, which is easier to invert.

Per-weight learning rates can be seen as a crude approximation of the diagonal Newton method. The number of true groundings not satisfied by evidence is a heuristic approximation to the count variance, which the diagonal Newton method uses to rescale each dimension of the gradient. The diagonal Newton method, however, can adapt to changes in the second derivative at different points in the weight space. Its main limitation is that clauses can be far from uncorrelated. The next method addresses this issue.

Scaled Conjugate Gradient

Gradient descent can be sped up by, instead of taking a small step of constant size at each iteration, performing a line search to find the optimum along the chosen descent direction. However, on ill-conditioned problems this is still inefficient because line searches along successive directions tend to partly undo the effect of each other: each line search makes the gradient along its direction zero, but the next line search will generally make it non-zero again. In long narrow valleys, instead of moving quickly to the optimum, gradient descent zigzags (see Figure 4.2).

A solution to this is to impose at each step the condition that the gradient along previous directions remain zero. The directions chosen in this way are called *conjugate*, and the method *conjugate gradient* [132]. Figure 4.2 shows an example of conjugate gradient on a quadratic surface. We use the *Polak-Ribiere* method for choosing conjugate gradients since it has generally been found to be the best-performing one. Conjugate gradient methods are some of the most efficient available, on a par with quasi-Newton ones. Unfortunately, applying them to MLNs is difficult, because line searches require computing the objective function, and therefore the partition function Z, which is highly intractable. (Computing Z is equivalent to computing all moments of the MLN, of which the gradient and Hessian are the first two.)

Fortunately, we can use the Hessian instead of a line search to choose a step size. This method is known as *scaled conjugate gradient* (SCG), and was originally proposed by Møller [86] for training neural networks. In our implementation, we choose a step size in the same way as diagonal Newton.

Conjugate gradient is usually more effective with a preconditioner, a linear transformation that attempts to reduce the condition number of the problem (e.g., [131]). Good precondition-ers approximate the inverse Hessian. We use the inverse diagonal Hessian as our preconditioner.

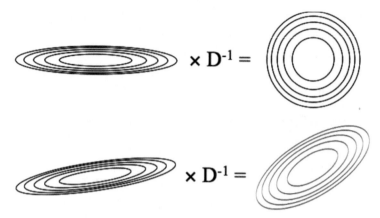

Figure 4.3: Application of the inverse diagonal Hessian preconditioner to two ill-conditioned learning problems. The surfaces of the objective functions are represented by their level sets. When the problem dimensions are uncorrelated, the final condition number will be 1, as in the upper problem.

Figure 4.3 shows how the inverse diagonal Hessian rescales problems to make them less skewed, reducing their condition number. Performance with the preconditioner is much better than without.

Of these methods, scaled conjugate gradient with a preconditioner is typically the most effective, but diagonal Newton also performs very well. See Lowd and Domingos [75] for more details and results.

4.2 STRUCTURE LEARNING AND THEORY REVISION

The structure of a Markov logic network is the set of formulas or clauses to which we attach weights. In principle, this structure can be learned or revised using any inductive logic programming (ILP) technique. However, since an MLN represents a probability distribution, much better results are obtained by using an evaluation function based on pseudo-likelihood, rather than typical ILP ones like accuracy and coverage [57]. Log-likelihood or conditional log-likelihood are potentially better evaluation functions, but are much more expensive to compute. In this section, we describe two algorithms for optimizing pseudo-likelihood with an ILP-style structure search. In experiments on real-world datasets, MLN structure learning algorithms have found better MLN rules than CLAUDIEN [22], FOIL [113], Aleph [140], and a hand-written knowledge base.

TOP-DOWN STRUCTURE LEARNING
The top-down structure learning (TDSL) methods of Kok and Domingos [57] learn or revise an MLN one clause at a time, one literal at a time. The initial structure can either be an empty network or existing KB. Either way, it is useful to start by adding all unit clauses (single atoms) to the MLN.

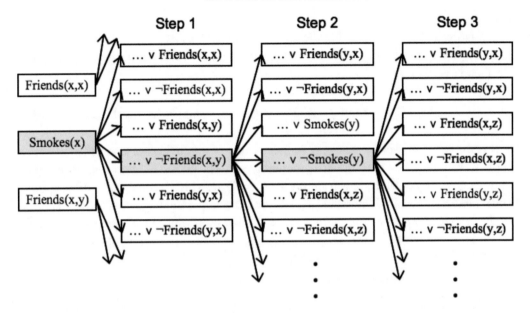

Figure 4.4: Top-down structure learning (TDSL) of a single clause through beam search when the beam width is one. In Step 1, TDSL extends all unit clauses with all possible literals and select the highest scoring candidate, Smokes(x) ∨ ¬Friends(x, y). (For space, we only show extensions of the Smokes(x) unit clause.) This process repeats with the new clause until no additional literal improves the score. Selected literals are shown in gray.

The weights of these capture (roughly speaking) the marginal distributions of the atoms, allowing the longer clauses to focus on modeling atom dependencies. To extend this initial model, TDSL either repeatedly finds the best clause using beam search and adds it to the MLN, or adds all "good" clauses of length l before trying clauses of length $l + 1$. Candidate clauses are formed by adding each predicate (negated or otherwise) to each current clause, with all possible combinations of variables, subject to the constraint that at least one variable in the new predicate must appear in the current clause. Hand-coded clauses are also modified by removing predicates.

Figure 4.4 shows an example of how TDSL can induce the clause Smokes(x) ∨ ¬Friends(x, y) ∨ ¬Smokes(y) (i.e., friends have similar smoking habits) through beam search. For simplicity, this example starts with the unit clauses and only considers adding literals.

We now discuss the evaluation measure, clause construction operators, and search strategy in greater detail.

Evaluation

As an evaluation measure, pseudo-likelihood (Equation 4.6) tends to give undue weight to the highest-arity predicates, resulting in poor modeling of the rest. We thus define the weighted pseudo-log-likelihood (WPLL) as

$$\log P_w^\bullet(X=x) = \sum_{r \in R} c_r \sum_{k=1}^{g_r} \log P_w(X_{r,k}=x_{r,k}|MB_x(X_{r,k})) \qquad (4.18)$$

where R is the set of first-order atoms, g_r is the number of groundings of first-order atom r, and $x_{r,k}$ is the truth value (0 or 1) of the kth grounding of r. The choice of atom weights c_r depends on the user's goals. By default, we can simply set $c_r = 1/g_r$, which has the effect of weighting all first-order predicates equally. If modeling a predicate is not important (e.g., because it will always be part of the evidence), we set its weight to zero. To combat overfitting, TDSL penalizes the WPLL with a structure prior of $e^{-\alpha \sum_{i=1}^{F} d_i}$, where d_i is the number of literals that differ between the current version of the clause and the original one. (If the clause is new, this is simply its length.) This is similar to the approach used in learning Bayesian networks [45].

A potentially serious problem that arises when evaluating candidate clauses using WPLL is that the optimal (maximum WPLL) weights need to be computed for each candidate. Given that this involves numerical optimization, and may need to be done thousands or millions of times, it could easily make the algorithm too slow to be practical. TDSL avoids this bottleneck by simply initializing L-BFGS with the current weights (and zero weight for a new clause) and relaxing its convergence threshold when evaluating candidate clauses. Second-order, quadratic-convergence methods such as L-BFGS are known to be very fast if started near the optimum. This is what happens in our case; L-BFGS typically converges in just a few iterations, sometimes one. The time required to evaluate a clause is in fact dominated by the time required to compute the number of its true groundings in the data. This time can be greatly reduced using sampling and other techniques [57].

Operators

When learning an MLN from scratch (i.e., from a set of unit clauses), the natural operator to use is the addition of a literal to a clause. When refining a hand-coded KB, the goal is to correct the errors made by the human experts. These errors include omitting conditions from rules and including spurious ones, and can be corrected by operators that add and remove literals from a clause. These are the basic operators used by TDSL. In addition, we have found that many common errors (wrong direction of implication, wrong use of connectives with quantifiers, etc.) can be corrected at the clause level by flipping the signs of atoms, and TDSL also allows this. When adding a literal to a clause, TDSL considers all possible ways in which the literal's variables can be shared with existing ones, subject to the constraint that the new literal must contain at least one variable that appears in an existing one. To control the size of the search space, TDSL limits the number of distinct variables in a clause. TDSL only removes literals from the original hand-coded clauses or their descendants, and only considers removing a literal if it leaves at least one path of shared variables between each pair of remaining literals.

Search

There are two alternative search strategies, one faster and one more complete. The first approach adds clauses to the MLN one at a time, using beam search to find the best clause to add: starting with the unit clauses and the expert-supplied ones, TDSL applies each legal literal addition and deletion to each clause, keeps the b best ones, applies the operators to those, and repeats until no new clause improves the WPLL. The chosen clause is the one with highest WPLL found in any iteration of the search. If the new clause is a refinement of a hand-coded one, it replaces it. (Notice that, even though TDSL both adds and deletes literals, no loops can occur because each change must improve WPLL to be accepted.)

The second approach adds k clauses at a time to the MLN, and is similar to that of McCallum for learning CRFs [78]. In contrast to beam search, which adds the best clause of any length found, this approach adds all "good" clauses of length l before attempting any of length $l + 1$. We call it *shortest-first search*. Shortest-first search is usually more expensive than beam search, but it tends to produce smaller and more accurate MLNs.

BOTTOM-UP STRUCTURE LEARNING

Top-down approaches, such as the one described in the previous section, follow a blind generate-and-test strategy in which many potential changes to an existing model are systematically generated independent of the training data, and then tested for empirical adequacy. For complex models such as MLNs, the space of potential revisions is combinatorially explosive and such a search can become difficult to control, resulting in convergence to suboptimal local maxima. Bottom-up learning methods attempt to use the training data to directly construct promising structural changes or additions to the model, avoiding many of the local maxima and plateaus in a large search space [88].

The BUSL (bottom-up structure learning) algorithm, introduced by Mihalkova and Mooney [83], applies this bottom-up approach to the task of learning MLN structure. Empirically, BUSL often yields more accurate models much faster than purely top-down structure learning.

BUSL uses a propositional Markov network structure learner to construct template networks that then guide the construction of candidate clauses. The template networks are composed of *template nodes*, conjunctions of one or more literals that serve as building blocks for creating clauses. Template nodes are constructed by looking for groups of constant-sharing ground literals that are true in the data and abstracting them by substituting variables for the constants. Thus, these template nodes could also be viewed as portions of clauses that have true groundings in the data. To understand why conjunctions of literals with true groundings are good candidates for clause components, consider the special case of a definite clause: $L_1 \wedge \cdots \wedge L_n \Rightarrow P$. If the conjoined literals in the body have no true groundings, then the clause is always trivially satisfied. Therefore, we expect that true conjunctions will be most useful for building effective clauses. (Note that the conjunction in the antecedent becomes a disjunction when we rewrite the implication in normal form.) The template Markov networks encode independencies among these template nodes. A template network does not specify the actual MLN clauses or their weights. To search for these, BUSL generates clause candidates by focusing on each maximal clique in turn and producing all possible clauses consistent with it. The candidates are then evaluated using the WPLL score, as in top-down structure learning.

We now discuss how BUSL builds template nodes, template networks, and candidate clauses in greater detail. A separate template and set of template nodes is created for each predicate P. First, BUSL creates a *head* template node consisting of a literal for P in which each argument is assigned a unique variable. This template node is analogous to the head in a definite clause; however, note that BUSL is not limited to constructing definite clauses. Starting with each ground literal l of predicate P, BUSL finds "chains" of literals connected by shared constant arguments. Each chain is abstracted by replacing constants with variables. The resulting conjunctions of "variablized" literals become the template nodes. BUSL keeps track of which template nodes can be created from each ground literal l of predicate P. These statistics are used to create a set of examples – one example per literal, each with one attribute per template node – for training the template Markov network. Mihalkova and Mooney chose the recent Grow-Shrink Markov Network (GSMN) algorithm by Bromberg *et al.* [13]. GSMN uses χ^2 statistical tests to determine whether two nodes are conditionally independent of each other.

The resulting Markov network captures independencies among the template nodes: each template node is independent of all others given its immediate neighbors (i.e. its Markov blanket). Because of these independencies, BUSL can restrict its structure search for clauses only to those candidates whose literals correspond to template nodes that form a clique in the template. BUSL makes a number of additional restrictions on the search in order to decrease the number of free variables in a clause, thus decreasing the size of the ground MLN during inference, and further reducing the search space. Complying with these restrictions, BUSL considers each clique in which the head template node participates and constructs all possible clauses of length one to the size of the clique by forming disjunctions from the literals of the participating template nodes with all possible negation/non-negation combinations.

After template creation and clause candidate generation are carried out for each predicate in the domain, duplicates are removed and the candidates are evaluated using the WPLL. The WPLL scoring is done by generatively learning weights for all clauses at once with L-BFGS. After all candidates are scored, they are considered for addition to the MLN in order of decreasing score. To reduce overfitting and speed up inference, only candidates with weight greater than a fixed threshold are considered. Candidates that do not increase the overall WPLL of the learned structure are discarded.

See Mihalkova and Mooney [83] for additional details, pseudocode, and examples of how BUSL works.

4.3   UNSUPERVISED LEARNING

In supervised learning, the true values of all predicates are known in the training data; in unsupervised learning, this is no longer the case. We will begin by discussing the simple case of missing data, where some or all groundings of a predicate are unknown in the training data. Partial missing data is common when some attributes are more expensive to obtain than others. Model-based clustering is an example of an entirely unknown predicate: the cluster assignment for each example is never observed, but a good model and clustering can be found together with the expectation maximization (EM) algorithm. In these examples, something is always known about the predicate – either the values of some of its groundings or the structure of its relationships with other predicates. In the second part of this section, we will discuss statistical predicate invention, in which the goal is to discover new concepts that help explain the observed data with minimal guidance or restrictions on what those concepts could represent. As in the supervised case, we optimize our models by maximizing the likelihood of the observed data; the key difference is that unsupervised models must sum out some variables to obtain this likelihood.

WEIGHT LEARNING

Suppose that some subset Y of the ground atoms have unknown values in the training data. The likelihood of the observed atoms, X, can be found by summing out the unobserved ones:

$$P_w(X=x) = \sum_y P_w(X=x, Y=y) = \sum_y \frac{1}{Z} \exp\left(\sum_i w_i n_i(x, y)\right) \qquad (4.19)$$

The gradient of the log likelihood is analogous to the fully observed case, except that it is now a difference of expectations:

$$\frac{\partial}{\partial w_i} \log P_w(x) = \sum_{y'} P_w(x, y') n_i(x, y') - \sum_{x', y'} P_w(x', y') n_i(x', y')$$
$$= E_{w,y}[n_i(x, y)] - E_{w,x,y}[n_i(x, y)] \qquad (4.20)$$

where the expectation $E_{w,y}$ is over the unknown atoms, Y, and $E_{w,y,x}$ is over all atoms, $Y \cup X$. (Both expectations are computed according to the current model, P_w.) Calculating these expectations is typically intractable, but we can choose to optimize the pseudo-likelihood instead. The key difference from weight learning with complete data is that we must run inference to estimate the state of Y even when using pseudo-likelihood.

We can do discriminative weight learning with missing data as well. There are now three classes of literals in the domain: known evidence atoms, X; observed query atoms, Y_o; and unobserved query atoms, Y_u (note the change in notation from the previous paragraph). Since X is evidence and Y_u is unknown, the only likelihood we can optimize is that of Y_o. Our ability to learn a meaningful model therefore depends on having a non-empty set of observed query atoms, Y_o. The gradient is the same as the generative case except that we are now conditioning on the evidence atoms x:

$$\frac{\partial}{\partial w_i} \log P_w(y_o|x) = \sum_{y_u'} P_w(y_o, y_u'|x) n_i(x, y_o, y_u') - \sum_{y_o', y_u'} P_w(y_o', y_u'|x) n_i(x, y_o', y_u')$$
$$= E_{w,y_u}[n_i(x, y_o, y_u)] - E_{w,y_o,y_u}[n_i(x, y_o, y_u)] \qquad (4.21)$$

The Hessian matrix is now a difference of covariance matrices: the covariance matrix when only unknown query atoms vary minus the covariance matrix when all query atoms vary. As in the fully observed case, the gradient and Hessian are intractable to compute but can be approximated using MCMC. The same types of numeric optimization methods also apply, with one caveat: the (conditional) log likelihood may no longer be convex when there is missing data. Therefore, when using scaled conjugate gradient, we recommend being more conservative in adjusting the size of the trust region that limits step size. For more information on unsupervised weight learning and an application to coreference resolution, see Poon and Domingos [107].

Missing data can also be handled by the expectation maximization (EM) algorithm [25]. EM splits parameter optimization into two steps: the expectation step (E-step) fills in missing data using the current model and the maximization step (M-step) adjusts the model parameters to maximize the likelihood of the filled-in data. Repeating these two steps is guaranteed to lead to a local optimum. In many applications of EM, such as mixture models, the M-step is trivial. For Markov logic, however, selecting the maximum likelihood parameters is typically very hard – attempting a full weight optimization in every iteration of EM would be very expensive, and almost certainly a waste of effort. Also, in MLNs in general, there is a single, large, interconnected learning example, and so the M step is not amortized over many independent examples. Thus directly optimizing the expected likelihood using gradient descent (or second-order techniques) is likely to be preferable to EM.

STATISTICAL PREDICATE INVENTION

Statistical predicate invention is the discovery of new concepts, properties and relations from data, expressed in terms of the observable ones, using statistical techniques to guide the process and explicitly representing the uncertainty in the discovered predicates. These can in turn be used as a basis for discovering new predicates, which is potentially much more powerful than learning based on a fixed set of simple primitives. Essentially all the concepts used by humans can be viewed as invented predicates, with many levels of discovery between them and the sensory percepts they are ultimately based on.

In statistical learning, this problem is known as hidden or latent variable discovery, and in relational learning as predicate invention. Both hidden variable discovery and predicate invention are considered quite important in their respective communities, but are also very difficult, with limited progress to date.

Only a few approaches combine elements of statistical and relational learning. Most of them learn a single relation or perform a single clustering [110, 153, 91, 155, 74, 124, 16, 21]. The most advanced approach is the MRC (Multiple Relational Clusterings) algorithm by Kok and Domingos [58]. MRC automatically invents predicates by clustering objects, attributes and relations. MRC learns multiple clusterings, rather than just one, to represent the complexities in relational data. The invented predicates capture arbitrary regularities over all relations, and are not just used to predict a designated target relation.

MRC is based on the observation that, in relational domains, multiple clusterings are necessary to fully capture the interactions between objects. Consider the following simple example. People have

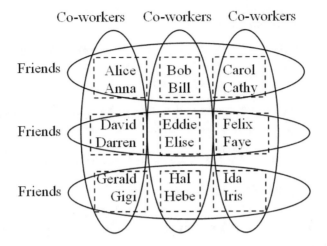

Figure 4.5: Example of multiple clusterings. Friends are clustered together in horizontal ovals, and co-workers are clustered in vertical ovals.

coworkers, friends, technical skills, and hobbies. People's technical skills are best predicted by their coworkers' skills, and their hobbies by their friends' hobbies. If we form a single clustering of people, coworkers and friends will be mixed, and our ability to predict both skills and hobbies will be hurt. Instead, we should cluster together people who work together, and simultaneously cluster people who are friends with each other. Each person thus belongs to both a "work cluster" and a "friendship cluster." (See Figure 4.5.) Membership in a work cluster is highly predictive of technical skills, and membership in a friendship cluster is highly predictive of hobbies. The remainder of this section presents a formalization of this idea and an efficient algorithm to implement it.

Notice that multiple clusterings may also be useful in propositional domains, but the need for them there is less acute because objects tend to have many fewer properties than relations. (For example, Friends(Anna, x) can have as many groundings as there are people in the world, and different friendships may be best predicted by different clusters Anna belongs to.)

The statistical model used by MRC is defined using finite second-order Markov logic, in which variables can range over relations (predicates) as well as objects (constants). Extending Markov logic to second-order involves simply grounding atoms with all possible predicate symbols as well as all constant symbols, and allows us to represent some models much more compactly than first-order Markov logic. We use it to specify how predicate symbols are clustered.

We use the variable r to range over predicate symbols, x_i for the ith argument of a predicate, γ_i for a cluster of ith arguments of a predicate (i.e., a set of symbols), and Γ for a clustering (i.e., a set of clusters or, equivalently, a partitioning of a set of symbols). For simplicity, we present our rules in generic form for predicates of all arities and argument types, with n representing the arity of r;

in reality, if a rule involves quantification over predicate variables, a separate version of the rule is required for each arity and argument type.

The first rule in our second-order MLN for predicate invention states that each symbol belongs to at least one cluster:

$$\forall x \: \exists \gamma \: x \in \gamma$$

This rule is hard, i.e., it has infinite weight and cannot be violated. The second rule states that a symbol cannot belong to more than one cluster in the same clustering:

$$\forall x, \gamma, \gamma', \Gamma \: \: x \in \gamma \land \gamma \in \Gamma \land \gamma' \in \Gamma \land \gamma \neq \gamma' \Rightarrow x \notin \gamma'$$

This rule is also hard. It is called the *mutual exclusion* rule.

If r is in cluster γ_r and x_i is in cluster γ_i, we say that $r(x_1, \ldots, x_n)$ is in the *combination of clusters* $(\gamma_r, \gamma_1, \ldots, \gamma_n)$. The next rule says that each atom appears in exactly one combination of clusters, and is also hard:

$$\forall r, x_1, \ldots, x_n \: \exists^! \gamma_r, \gamma_1, \ldots, \gamma_n$$
$$r \in \gamma_r \land x_1 \in \gamma_1 \land \ldots \land x_n \in \gamma_n$$

(The $\exists^!$ quantifier is used here to mean "exactly one combination exists.") The next rule is the key rule in the model, and states that the truth value of an atom is determined by the cluster combination it belongs to:

$$\forall r, x_1, \ldots, x_n, +\gamma_r, +\gamma_1, \ldots, +\gamma_n$$
$$r \in \gamma_r \land x_1 \in \gamma_1 \land \ldots \land x_n \in \gamma_n \Rightarrow r(x_1, \ldots, x_n)$$

This rule is soft. The "+" notation is syntactic sugar that signifies the MLN contains an instance of this rule *with a separate weight* for each tuple of clusters $(\gamma_r, \gamma_1, \ldots, \gamma_n)$. As we will see below, this weight is the log-odds that a random atom in this cluster combination is true. Thus, this is the rule that allows us to predict the probability of query atoms given the cluster memberships of the symbols in them. We call this the *atom prediction* rule. Combined with the mutual exclusion rule, it also allows us to predict the cluster membership of evidence atoms. Chaining these two inferences allows us to predict the probability of query atoms given evidence atoms.

To combat the proliferation of clusters and consequent overfitting, we impose an exponential prior on the number of clusters, represented by the formula

$$\forall \gamma \: \exists x \: x \in \gamma$$

with negative weight $-\lambda$. The parameter λ is fixed during learning, and is the penalty in log-posterior incurred by adding a cluster to the model. Thus larger λs lead to fewer clusterings being formed.[2]

A *cluster assignment* $\{\Gamma\}$ is an assignment of truth values to all $r \in \gamma_r$ and $x_i \in \gamma_i$ atoms. The MLN defined by the five rules above represents a joint distribution $P(\{\Gamma\}, R)$ over $\{\Gamma\}$ and R, the

---

[2] We have also experimented with using a Chinese restaurant process prior (CRP, [102]), and the results were similar. We thus use the simpler exponential prior.

vector of truth assignments to the observable ground atoms. Learning consists of finding the cluster assignment that maximizes $P(\{\Gamma\}|R) \propto P(\{\Gamma\}, R) = P(\{\Gamma\})P(R|\{\Gamma\})$, and the corresponding weights. $P(\{\Gamma\}) = 0$ for any $\{\Gamma\}$ that violates a hard rule. For the remainder, $P(\{\Gamma\})$ reduces to the exponential prior. It is easily seen that, given a cluster assignment, the MLN decomposes into a separate MLN for each combination of clusters, and the weight of the corresponding atom prediction rule is the log odds of an atom in that combination of clusters being true. (Recall that, by design, each atom appears in exactly one combination of clusters.) Further, given a cluster assignment, atoms with unknown truth values do not affect the estimation of weights, because they are graph-separated from all other atoms by the cluster assignment. If t_k is the empirical number of true atoms in cluster combination k, and f_k the number of false atoms, we estimate w_k as $\log((t_k + \beta)/(f_k + \beta))$, where β is a smoothing parameter.

Conversely, given the model weights, we can use inference to assign probabilities of membership in combinations of clusters to all atoms. Thus the learning problem can in principle be solved using an EM algorithm, with cluster assignment as the E step, and MAP estimation of weights as the M step. We begin by simplifying the problem by performing hard assignment of symbols to clusters (i.e., instead of computing probabilities of cluster membership, a symbol is simply assigned to its most likely cluster). Since performing an exhaustive search over cluster assignments is infeasible, the key is to develop an intelligent tractable approach. This makes the M step trivial, but the E step is still extremely complex. Since, given a cluster assignment, the MAP weights can be computed in closed form, a better alternative to EM is simply to search over cluster assignments, evaluating each assignment by its posterior probability. This can be viewed as a form of structure learning, where a structure is a cluster assignment.

We now describe the MRC algorithm. The basic idea is that, when clustering sets of symbols related by atoms, each refinement of one set of symbols potentially forms a basis for the further refinement of the related clusters. MRC is thus composed of two levels of search: the top level finds clusterings, and the bottom level finds clusters. At the top level, MRC is a recursive procedure whose inputs are a cluster of predicates γ_r per arity and argument type, and a cluster of symbols γ_i per type. In the initial call to MRC, each γ_r is the set of all predicate symbols with the same number and type of arguments, and γ_i is the set of all constant symbols of the ith type. At each step, MRC creates a cluster symbol for each cluster of predicate and constant symbols it receives as input. Next it clusters the predicate and constant symbols, creating and deleting cluster symbols as it creates and destroys clusters. It then calls itself recursively with each possible combination of the clusters it formed. For example, suppose the data consists of binary predicates $r(x_1, x_2)$, where x_1 and x_2 are of different type. If r is clustered into γ_r^1 and γ_r^2, x_1 into x_1^1 and x_1^2, and x_2 into x_2^1 and x_2^2, MRC calls itself recursively with the cluster combinations $(\gamma_r^1, \gamma_1^1, \gamma_2^1)$, $(\gamma_r^1, \gamma_1^1, \gamma_2^2)$, $(\gamma_r^1, \gamma_1^2, \gamma_2^1)$, $(\gamma_r^1, \gamma_1^2, \gamma_2^2)$, $(\gamma_r^2, \gamma_1^1, \gamma_2^1)$, etc.

Within each recursive call, MRC uses greedy search with restarts to find the MAP clustering of the subset of predicate and constant symbols it received. It begins by assigning all constant symbols of the same type to a single cluster, and similarly for predicate symbols of the same arity and argument type. The search operators used are: move a symbol between clusters, merge two clusters, and split a cluster. (If clusters are large, only a random subset of the splits is tried at each step.) A greedy search ends when no operator increases posterior probability. Restarts are performed, and they give different results because of the random split operator used.

Notice that in each call of MRC, it forms a clustering for each of its input clusters, thereby always satisfying the first two hard rules in the MLN. MRC also always satisfies the third hard rule because it only passes the atoms in the current combination to each recursive call.

MRC terminates when no further refinement increases posterior probability, and returns the finest clusterings produced. In other words, if we view MRC as growing a tree of clusterings, it returns the leaves. Conceivably, it might be useful to retain the whole tree, and perform shrinkage [79] over it. This is an item for future work. Notice that the clusters created at a higher level of recursion constrain the clusters that can be created at lower levels, e.g., if two symbols are assigned to different clusters at a higher level, they cannot be assigned to the same cluster in subsequent levels. Notice also that predicate symbols of different arities and argument types are never clustered together. This is a limitation that we plan to overcome in the future.

Statistical predicate invention can be seen as an alternative to structure learning in relational domains. Traditional relational structure learning approaches such as ILP build formulas by incrementally adding predicates that share variables with existing predicates. The dependencies these formulas represent can also be captured by inventing new predicates. For example, consider a formula that states that if two people are friends, either both smoke or neither does: $\forall x \forall y \, Fr(x, y) \Rightarrow (Sm(x) \Leftrightarrow Sm(y))$. We can compactly represent this using two clusters, one containing friends who smoke, and one containing friends who do not. The model introduced in this section represents a first step in this direction.

4.4   TRANSFER LEARNING

The learning algorithms discussed so far work within a single domain: the only information used is contained within the dataset or manually specified (e.g., formulas, algorithm parameters, etc.). This approach overlooks a potentially valuable source of information: related domains with similar characteristics. Transfer learning is the task of automatically incorporating knowledge learned in these related domains to improve learning speed or accuracy in a new domain of interest. Interest in transfer learning has grown rapidly in recent years (e.g., [142, 31, 67, 125, 143, 144]). Work to date falls mainly into what may be termed *shallow transfer*: generalizing to different distributions over the same variables, or different variations of the same domain (e.g., different numbers of objects). What remains largely unaddressed is *deep transfer*: generalizing across domains (i.e., between domains where the types of objects and variables are different). In this section, we describe two recent methods for deep transfer learning in relational domains using Markov logic.

DEEP TRANSFER VIA AUTOMATIC MAPPING AND REVISION

Problems that are superficially different often contain analogous concepts. For example, directors working with actors on movies are in some ways similar to professors working with students on publications. Therefore, rules that apply to directors, actors, and movies may also apply to professors, students, and publications. The TAMAR (Transfer via Automatic Mapping and Revision) algorithm developed by Mihalkova *et al.* [82] finds and adapts these analogies automatically using Markov logic. Results in several real-world domains demonstrate that this approach successfully reduces the amount of time and training data needed to learn an accurate model of a target domain over learning from scratch.

TAMAR begins by learning an MLN for the source domain using structure learning methods such as those described in Section 4.2. For each learned rule, it finds the best mapping to predicates in the target domain, scored by WPLL (Equation 4.18). Once these rules have been mapped into the target domain, they are lengthened to become more specific or shortened to become more general. Finally, a relational pathfinding algorithm is used to discover additional rules. We next describe these steps in more detail.

Predicate Mapping

The first step is to find the best way to map a source MLN into a target MLN. The quality of a mapping is measured by the performance of the mapped MLN on the target data, as estimated by the WPLL score. The number of possible predicate mappings is exponential in the number of predicates. This makes finding the best global mapping computationally prohibitive in most cases. Instead, TAMAR finds the best local mapping for each clause, independent of how other clauses were mapped.

TAMAR does this through exhaustive search through the space of all legal mappings. A mapping is legal if each source predicate in a given clause is mapped either to a compatible target predicate or to the "empty" predicate, which erases all literals of that source predicate from the clause. Two predicates are compatible if they have the same arity and the types of their arguments are compatible with any existing type constraints. For any legal mapping, a type in the source domain is mapped to at most one corresponding type in the target domain, and the type constraints are formed by requiring that these type mappings be consistent across all predicates in the clause. After a legal mapping is established, TAMAR evaluates the translated clause by calculating the WPLL of an MLN consisting of only this translated clause. The best predicate mapping for a clause is the one whose translated clause has the highest WPLL score. Figure 4.6 shows the best predicate mapping found by the algorithm for a given source clause, when transferring from an academic domain to a film production one.

Note that because the predicate mapping algorithm may sometimes choose to map a predicate to an "empty" predicate, the entire structure is not necessarily always mapped to the target domain.

The mapped structure is then revised to improve its fit to the data. The skeleton of the revision algorithm has three steps and is similar to that of FORTE [115], which revises first-order theories.

Source clause:

$$Publication(T, A) \quad \wedge \quad Publication(T, B) \wedge Professor(A) \wedge Student(B)$$
$$\wedge \quad \neg SamePerson(A, B) \Rightarrow AdvisedBy(B, A)$$

Best mapping:

$$\begin{array}{rcl}
Publication(title, person) & \rightarrow & MovieMember(movie, person) \\
Professor(person) & \rightarrow & Director(person) \\
Student(person) & \rightarrow & Actor(person) \\
SamePerson(person, person) & \rightarrow & SamePerson(person, person) \\
AdvisedBy(person, person) & \rightarrow & WorkedFor(person, person)
\end{array}$$

Figure 4.6: An output of the predicate mapping algorithm.

1. **Self-Diagnosis:** The purpose of this step is to focus the search for revisions only on the inaccurate parts of the MLN. The algorithm inspects the source MLN and determines for each clause whether it should be shortened, lengthened, or left as is. For each clause C, this is done by considering every possible way of viewing C as an implication in which one of the literals is placed on the right-hand side of the implication and is treated as the conclusion and the remaining literals serve as the antecedents. The conclusion of a clause is drawn only if the antecedents are satisfied and the clause "fires." Thus, if a clause makes the wrong conclusion, it is considered for lengthening because the addition of more literals, or conditions, to the antecedent will make it harder to satisfy, thus preventing the clause from firing. On the other hand, there may be clauses that fail to draw the correct conclusion because there are too many conditions in the antecedents that prevent them from firing. In this case, we consider shortening the clause.

2. **Structure Update:** Clauses marked as too long are shortened, while those marked as too short are lengthened. This is done via beam search, using the WPLL for scoring candidate clauses.

3. **New Clause Discovery:** Relational pathfinding (RPF) [114] finds new clauses in the target domain. RPF is a bottom-up ILP technique similar to some of the methods used in BUSL (Section 4.2). Only clauses that improve the WPLL are added to the MLN. RPF and the previous structure update step operate independently of each other; in particular, the clauses discovered by RPF are not diagnosed or revised. However, better results are obtained if the clauses discovered by RPF are added to the MLN before carrying out the revisions.

DEEP TRANSFER VIA SECOND-ORDER MARKOV LOGIC

Some concepts, such as transitivity, are common to many different domains. Learning these concepts and looking for them in new domains can lead to faster and more accurate structure learning.

This is the approach used by DTM (Deep Transfer via Markov logic), introduced by Davis and Domingos [20]. Instead of constructing a specific mapping between source and target domains as TAMAR does, DTM extracts general, transferable knowledge that could be applied to any target domain. DTM has been successfully used to transfer learned knowledge among social network, molecular biology and Web domains. In addition to improving empirical performance over normal structure learning, DTM discovered patterns including broadly useful properties of predicates, like symmetry and transitivity, and relations among predicates, such as homophily.

In order to abstract away superficial domain descriptions, DTM uses second-order Markov logic, where formulas contain predicate variables to model common structures among first-order clauses. To illustrate the intuition behind DTM, consider the following two formulas:

$$\texttt{Complex(z, y)} \wedge \texttt{Interacts(x, z)} \Rightarrow \texttt{Complex(x, y)}$$
$$\texttt{Location(z, y)} \wedge \texttt{Interacts(x, z)} \Rightarrow \texttt{Location(x, y)}$$

Both are instantiations of the following clause:

$$\texttt{r(z, y)} \wedge \texttt{s(x, z)} \Rightarrow \texttt{r(x, y)}$$

where r and s are predicate variables. Introducing predicate variables allows DTM to represent high-level structural regularities in a domain-independent fashion. This knowledge can be transferred to another problem, where the clauses are instantiated with the appropriate predicate names. The key assumption that DTM makes is that the target domain shares some second-order structure with the source domain.

Given a set of first-order formulas, DTM converts each formula into second-order logic by replacing all predicate names with predicate variables. It then groups the second-order formulas into cliques. Two second-order formulas are assigned to the same clique if and only if they are over the same set of literals. DTM evaluates which second-order cliques represent regularities whose probability deviates significantly from independence among their subcliques. It selects the top k highest-scoring second-order cliques to transfer to the target domain. The transferred knowledge provides a declarative bias for structure learning in the target domain.

The four key elements of DTM, introduced in the next subsections, are: (i) how to define cliques, (ii) how to search for cliques, (iii) how to score the cliques and (iv) how to apply cliques to a new problem.

Second-order Cliques

DTM uses second-order cliques to model second-order structure. It is preferable to use this representation as opposed to arbitrary second-order formulas because multiple different formulas over the same predicates can capture the same regularity. A clique groups those formulas with similar effects into one structure. A set of literals with predicate variables, such as $\{\texttt{r(x, y)}, \texttt{r(y, x)}\}$, defines each second-order clique and the states, or features, are conjunctions over the literals in the clique. It is more convenient to look at conjunctions than clauses as they do not overlap, and can be evaluated separately. The features correspond to all possible ways of negating the literals in a clique. The features for $\{r(x, y), r(y, x)\}$ are $\{r(x, y) \wedge r(y, x)\}$, $\{r(x, y) \wedge \neg r(y, x)\}$, $\{\neg r(x, y) \wedge r(y, x)\}$ and $\{\neg r(x, y) \wedge \neg r(y, x)\}$.

DTM imposes the following restrictions on cliques:

1. The literals in the clique are connected. That is, a path of shared variables must connect each pair of literals in the clique.

2. No cliques are the same modulo variable renaming. For example, $\{r(x, y), r(z, y), s(x, z)\}$ and $\{r(z, y), r(x, y), s(z, x)\}$, where r and s are predicate variables, are equivalent as the second clique renames variable x to z and variable z to x.

The states of a clique are all possible ways of negating the literals in the clique subject to the following constraints:

1. No two features are the same modulo variable renaming.

2. Two distinct variables are not allowed to unify. For example, $r(x, y) \wedge r(y, x)$ really represents the following formula: $r(x, y) \wedge r(y, x) \wedge x \neq y$. This constraint ensures that a possible grounding does not appear in two separate cliques. Without this constraint, all true groundings of $r(x, y) \wedge s(x, x)$ would also be true of $r(x, y) \wedge s(x, z)$ (which appears as a feature in a different clique).

Second-Order Clique Evaluation

Intuitively, the goal is to assign a high score to a second-order clique that captures a dependency between its literals. Each clique can be decomposed into pairs of subcliques, and it should capture a regularity beyond what its subcliques represent. For example, the second-order clique $\{r(x, y), r(z, y), s(x, z)\}$ can be decomposed into the following three pairs of subcliques: (i) $\{r(x, y), r(z, y)\}$ and $\{s(x, z)\}$, (ii) $\{r(x, y), s(x, z)\}$ and $\{r(z, y)\}$, and (iii) $\{r(z, y), s(x, z)\}$ and $\{r(x, y)\}$.

To score a second-order clique, each of its first-order instantiations is evaluated. To score a first-order clique, DTM checks if its probability distribution is significantly different from the product of the probabilities of each possible pair of subcliques that it can be decomposed into. The natural way to compare these two distributions is to use the Kullback-Leibler (K-L) divergence:

$$D(p\|q) = \sum_{x} p(x) \log \frac{p(x)}{q(x)} \tag{4.22}$$

where p is the clique's probability distribution, and q is the distribution it would have if the two subcliques were independent. DTM uses Bayesian estimates of the probabilities with Dirichlet priors with all $\alpha_i = 1$.

For each first-order instantiation of a second-order clique, DTM computes its K-L divergence versus all its decompositions. Each instantiation receives the minimum K-L score over the set of its decompositions, because any single one could explain the clique's distribution. Each second-order clique receives the average score of its top m first-order instantiations, in order to favor second-order cliques that have multiple useful instantiations.

Search

DTM works with any learner that induces formulas in first-order logic. We have explored two separate strategies for inducing formulas in the source domain: exhaustive search and beam search.

Exhaustive search. Given a source domain, the learner generates all first-order clauses up to a maximum clause length and a maximum number of object variables. The entire set of clauses is passed to DTM for evaluation.

Beam search. Exhaustive search does not scale well, as the number of clauses it produces is exponential in the clause length, and it becomes computationally infeasible to score long clauses. Beam search, a common strategy for scaling structure learning, is used in top-down structure learning (TDSL) (Section 4.2). However, transfer learning and structure learning have different objectives. In transfer learning, the goal is to derive a large, diverse set of clauses to evaluate for potential transfer to the target domain. Structure learning simply needs to induce a compact theory that accurately models the predicates in the source domain. The theories induced by TDSL tend to contain very few clauses and thus are not ideal for transfer. An alternative approach is to induce a separate theory to predict each predicate in the domain. However, the resulting theory may not be very diverse, since clauses will contain only the target predicate and predicates in its Markov blanket (i.e., its neighbors in the network). A better approach is to construct models that predict sets of predicates.

Given the final set of learned models, DTM groups the clauses into second-order cliques and evaluates each clique that appears more than twice.

Transfer Mechanism

The next question is how to make use of the transferred knowledge in the target domain. A key component of an inductive logic programming (ILP) [33] system is the declarative bias. Due to the large search space of possible first-order clauses, devising a good declarative bias is crucial for achieving good results with an ILP system. In ILP, two primary methods exist for expressing a declarative bias, and both forms of bias are often used in the same system. The first method restricts the search space. Common strategies include having type constraints, forcing certain predicate arguments to contain bound variables, and setting a maximum clause length. The second method involves incorporating background knowledge. Background knowledge comes in the form of hand-crafted clauses that define additional predicates, which can be added to a clause under construction. Effectively, background knowledge allows the learner to add multiple literals to a clause at once and overcome the myopia of greedy search. It is important to note that these common strategies can easily be expressed in second-order logic, and this is part of what motivates this approach. DTM can be viewed as a way to learn the declarative bias in one domain and apply it in another, as opposed to having a user hand-craft the bias for each domain.

When applying a second-order clique in a target domain, DTM decomposes the clique into a set of clauses, and transfers each clause. It uses clauses instead of conjunctions since most structure learning approaches, both in Markov logic and ILP, construct clauses. In Markov logic, a conjunction can be converted into an equivalent clause by negating each literal in it and flipping the sign of its weight.

There are three different ways to reapply the knowledge in the target domain:

Transfer by Seeding the Beam. In the first approach, the second-order cliques provide a declarative bias for the standard MLN structure search. DTM selects the top k clique templates that have at least one true grounding in the target domain. At the beginning of each iteration of the beam search, DTM seeds the beam with the clauses obtained from each legal instantiation of a clique in the target domain (i.e., only instantiations that conform with the type constraints of the domain are considered). This strategy forces certain clauses to be evaluated in the search process that would not be scored otherwise and helps overcome some of the limitations of greedy search.

Greedy Transfer without Refinement. The second approach again picks the top k clique templates that have at least one true grounding in the target domain and creates all legal instantiations in the target domain. This algorithm imposes a very stringent bias by performing a structure search where it only considers including the transferred clauses. The algorithm evaluates all clauses and greedily picks the one that leads to the biggest improvement in WPLL. The search terminates when no clause addition improves the WPLL.

Greedy Transfer with Refinement. The final approach adds a refinement step to the previous algorithm. In this case, the MLN generated by the greedy procedure serves as seed network during standard structure learning. The MLN search can now refine the clauses picked by the greedy procedure to better match the target domain. Additionally, it can induce new clauses to add to the theory.

In transfer learning experiments among social network, molecular biology, a Web domains, we found that DTM was more effective than top-down structure learning (which transfers no knowledge), and that greedy transfer with refinement was the most accurate DTM method [20]. In general, exhaustive search seems to yield better and more stable results than beam search, perhaps because beam search still focuses too heavily on finding a good theory as opposed to finding the best clauses for transfer.

FURTHER READING

The first learning algorithm for Markov logic was generative weight learning with pseudo-likelihood; see Richardson and Domingos for details and experimental results [117]. Pseudo-likelihood has also been used for structure learning by Kok and Domingos [57] and Mihalkova and Mooney [83].

Pseudo-likelihood was originally introduced by Besag [6]; for a theoretical comparison to exact generative and discriminative learning, see Liang and Jordan [68].

Discriminative weight learning in Markov logic was first done by Singla and Domingos [133] using a variant of Collins' voted perceptron algorithm [15]. See Lowd and Domingos [75] for details and an empirical comparison of all discriminative weight learning algorithms discussed in this chapter.

For more information on top-down and bottom-up structure learning, see Kok and Domingos [57] and Mihalkova and Mooney [83], respectively. Both approaches outperform directly using the output of various ILP algorithms, as previously done by Richardson and Domingos [117]. Most recently, Huynh and Mooney [50] introduced a method for discriminative structure and parameter learning. They examined the classic ILP scenario in which a single predicate is predicted given all others. Their method consisted of generating clauses using traditional ILP algorithms, selecting and weighting them using Markov logic weight learning with an L1 prior, and adding transitivity clauses as appropriate to allow for joint inference. Biba *et al.* [8] propose a discriminative structure learning algorithm that uses pseudo-likelihood to set weights and filter clause candidates, but does the final clause evaluation by CLL using MC-SAT. Instead of greedy or exhaustive search, the algorithm uses stochastic local search to find many locally optimal candidate clauses.

See Poon and Domingos [107] for an example of unsupervised weight learning. Details on statistical predicate invention can be found in Kok and Domingos [58]. For a large-scale application of these ideas to the Web, see the more recent Kok and Domingos publication [59] (also discussed in Chapter 6).

For transfer learning details and experiments, consult Mihalkova *et al.* [82] regarding TAMAR and Davis and Domingos [20] regarding DTM.

CHAPTER 5

Extensions

In this chapter, we describe how to overcome a number of limitations of the basic Markov logic representation introduced in Chapter 2. We extend Markov logic to handle continuous variables and infinite domains; uncertain disjunctions and existential quantifiers; and relational decision theory. We also describe inference algorithms for these extensions.

5.1 CONTINUOUS DOMAINS

In this section, we discuss how to extend Markov logic networks to hybrid domains involving both discrete and continuous variables, as developed by Wang and Domingos [148]. We call the generalized representation *hybrid Markov logic networks (HMLNs)*. We have developed efficient algorithms for inference in HMLNs, combining ideas from satisfiability testing, slice-sampling MCMC, and numerical optimization. Because most probabilistic models are easily translated into HMLNs, our algorithms are a powerful general-purpose tool for inference in structured domains. Weight learning algorithms are straightforward extensions of existing ones for MLNs. An application of HMLNs to a robot mapping domain appears in Chapter 6.

REPRESENTATION

Conceptually, extending MLNs to numeric and hybrid domains is quite straightforward: it suffices to allow numeric properties of objects as nodes, in addition to Boolean ones, and numeric terms as features, in addition to logical formulas. Since the syntax of first-order logic already includes numeric terms, no new constructs are required. A numeric term is a term of numeric type, and a numeric property is a designated function of numeric type. For example, if we are interested in distances between objects as random variables, we can introduce the numeric property `Distance(x, y)`.

Definition 5.1. A *hybrid Markov logic network* L is a set of pairs (F_i, w_i), where F_i is a first-order formula or a numeric term, and w_i is a real number. Together with a finite set of constants $C = \{c_1, c_2, \dots, c_{|C|}\}$, it defines a Markov network $M_{L,C}$ as follows:

1. $M_{L,C}$ contains one node for each possible grounding with constants in C of each predicate or numeric property appearing in L. The value of a predicate node is 1 if the ground predicate is true, and 0 otherwise. The value of a numeric node is the value of the corresponding ground term.

2. $M_{L,C}$ contains one feature for each possible grounding with constants in C of each formula or numeric term F_i in L. The value of a formula feature is 1 if the ground formula is true, and 0

otherwise. The value of a numeric feature is the value of the corresponding ground term. The weight of the feature is the w_i associated with F_i in L.

Thus an HMLN defines a family of log-linear models of the form $P(X=x) = \frac{1}{Z} \exp\left(\sum_i w_i s_i(x)\right)$, where s_i is the sum of the values of all groundings of F_i in x. Notice that logical formulas may contain numeric terms and these may contain indicator functions, allowing for arbitrary hybrid features. As in discrete MLNs, we assume that the value of every function for every tuple of arguments is known, and thus when grounding an HMLN every functional term can be replaced by its value. Proper measures and finite solutions are guaranteed by requiring that all variables have finite range, and features be finite everywhere in this range. For convenience, we allow some extensions of first-order syntax in defining HMLNs:

Infix notation. Numeric terms may appear in infix form.

Formulas as indicators. First-order formulas may be used as indicator functions within numeric terms.

Soft equality. $\alpha = \beta$ may be used as a shorthand for $-(\alpha - \beta)^2$, where α and β are arbitrary numeric terms. This makes it possible to state numeric constraints as equations, with an implied Gaussian penalty for diverging from them. If the weight of a formula is w, the standard deviation of the Gaussian is $\sigma = 1/\sqrt{2w}$. A numeric domain can now be modeled simply by writing down the equations that describe it. In our experience, this is the most common use of numeric features in HMLNs.

Soft inequality. $\alpha > t$ may be used as a shorthand for $-\log(1 + e^{a(t-\alpha)})$, and $\alpha < t$ for $-\log(1 + e^{a(\alpha-t)})$, with α an arbitrary numeric term. In other words, the (log) sigmoid function is used to represent soft inequality, with a controlling the degree of softness.

We now present algorithms for inference in HMLNs by extending the corresponding ones for MLNs. All algorithms assume logical formulas have been converted to standard clausal form. For simplicity, the exposition assumes that all weights are positive. Weights can be learned using the same algorithms as for MLNs, with feature counts generalized to feature sums. Other numeric parameters require a straightforward extension of these algorithms. HMLNs structure learning is a topic for future research.

INFERRING THE MOST PROBABLE STATE

The hybrid MaxWalkSAT (HMWS) algorithm performs MPE inference in HMLNs by combining MaxWalkSAT and L-BFGS [148]. Pseudocode for it is shown in Table 5.1, where **x** is the current state and $S(\mathbf{x})$ is the current sum of weighted features. Numeric variables are initialized uniformly at random, with user-determined range. At each search step, HMWS randomly chooses an unsatisfied clause or numeric feature c, and performs a random step with probability p and a greedy one otherwise. In random steps, HMWS chooses a variable at random and sets it to the value that maximizes $c(\mathbf{x})$. For a Boolean variable in an unsatisfied clause, this simply means flipping it. For a numeric variable, the maximization is done using L-BFGS, and HMWS then adds Gaussian noise to it (with user-determined variance). In greedy steps, HMWS first performs a one-dimensional optimization of $S(\mathbf{x})$ as a function of each variable, and chooses the variable change that yields the greatest improvement in $S(\mathbf{x})$. In numeric features, if all one-variable changes fail to improve S, HMWS performs a greedy or exhaustive search over assignments to the Boolean variables, for each assignment maximizing the feature sum as a function of the numeric variables using L-BFGS. The choice between greedy and exhaustive search is determined by the number of Boolean variables. The result of the numeric search for each setting of the Boolean variables is cached, to avoid redoing the search in the future. As in MaxWalkSAT, this process continues for a predefined number of steps, and is restarted a predefined number of times. The best assignment found is returned. When all variables and features are Boolean, HMWS reduces to MaxWalkSAT; when all are numeric, to a series of calls to L-BFGS.

INFERRING CONDITIONAL PROBABILITIES

Extending MC-SAT to handle numeric variables and features requires first extending WalkSAT and SampleSAT. A *Boolean constraint* is a clause. A *numeric constraint* is an inequality of the form $f_k(x) \geq a$. Given a set of constraints over Boolean and numeric variables, hybrid WalkSAT (HWS) attempts to find the assignment of values to the variables that maximizes the number of satisfied constraints. HWS is similar to HMWS with the functions $f_k(x)$ as numeric features, except that in each step maximization of $f_k(x)$ is halted as soon as $f_k(x) \geq a$, and the global objective function is the number of satisfied constraints, not the weighted sum of features.

Hybrid SampleSAT (HSS) generates a uniform sample from the states that satisfy a set of constraints M. It mixes HWS and simulated annealing steps. The energy (negative log probability) of a state for simulated annealing is the number of satisfied constraints. A new candidate state is generated by choosing a variable at random, flipping it if it is Boolean, and adding Gaussian noise with user-determined variance to it if it is numeric.

Hybrid MC-SAT (HMCS) is a slice sampler with one auxiliary variable per feature. In state x, the auxiliary variable u_k corresponding to feature f_k is sampled uniformly from $[0, e^{w_k f_k(x)}]$. For Boolean features, the construction of the constraint set M is the same as in MC-SAT. For numeric feature f_k, the standard slice sampling constraint $f_k \geq \log(u_k)/w_k$ is added to M. The constraints in M define the slice, and a new state is sampled from it using HSS. Pseudocode for HMCS is shown in Table 5.2, where the first step calls WalkSAT to satisfy all hard (infinite-weight) clauses, and $\mathcal{U}_S$ is the uniform distribution over set S. The proof that HMCS satisfies ergodicity and detailed balance is analogous to that for MC-SAT. In purely discrete domains, HMCS reduces to MC-SAT. In purely continuous ones, it is a new type of slice sampler, using a combination of simulated annealing and numerical optimization to very efficiently sample from the slice.

Table 5.1: MPE inference algorithm for hybrid MLNs.

---

function Hybrid-MaxWalkSAT(L, F, m_t, m_f, p)
    inputs: L, a set of weighted clauses
            F, a set of weighted numeric terms
            n_t, number of tries
            n_f, number of flips
            p, probability of a random move
    output: $\mathbf{x}^*$, best variable assignment found

$\mathbf{x}^* = \text{null}$
$S(\mathbf{x}^*) = -\infty$
for $i \leftarrow 1$ **to** n_t
  $\mathbf{x} \leftarrow$ random assignment
  for $j \leftarrow 1$ **to** n_f
    if $S(\mathbf{x}) > S(\mathbf{x}^*)$
      $\mathbf{x}^* \leftarrow \mathbf{x}$
      $S(\mathbf{x}^*) \leftarrow S(\mathbf{x})$
    $c \leftarrow$ a random unsatisfied clause or numeric term
    if uniform(0,1) $< p$
      select a random variable x appearing in c
      $x \leftarrow \text{argmax}_x c(\mathbf{x})$
      if x is numeric
        $x \leftarrow x + \text{Gaussian noise}$
    else
      for each variable x_i in c
        $x_i' \leftarrow \text{argmax}_{x_i} S(\mathbf{x})$
        $S(\mathbf{x}_i') \leftarrow S(\mathbf{x})$ with $x_i \leftarrow x_i'$
      $x_f' \leftarrow x_i'$ with highest $S(\mathbf{x}_i')$
      if $S(\mathbf{x}_f') > S(\mathbf{x})$ or c is a clause
        $x_f \leftarrow x_f'$
      else
        $\mathbf{x}_c \leftarrow \text{argmax}_{\mathbf{x}_c} S(\mathbf{x})$, where $\mathbf{x}_c$ is the subset of $\mathbf{x}$ appearing in c
return $\mathbf{x}^*$

Table 5.2: Efficient MCMC algorithm for hybrid MLNs.

---

function Hybrid-MC-SAT (L, F, n)
    inputs: L, a set of weighted clauses $\{(w_j, c_j)\}$
            F, a set of weighted numeric terms $\{(w_j, f_j)\}$
            n, number of samples
    output: $\{x^{(1)}, \ldots, x^{(n)}\}$, set of n samples

$x^{(0)} \leftarrow$ Satisfy(hard clauses in L)
for $i \leftarrow 1$ to n
    $M \leftarrow \emptyset$
    for all $(w_k, c_k) \in C$ satisfied by $x^{(i-1)}$
        with probability $1 - e^{-w_k}$ add c_k to M
    for all $(w_k, f_k) \in F$
    $u_k \sim \mathcal{U}_{[0, \exp(w_k f_k(x^{(i-1)}))]}$
    add $f_k(x^{(i)}) \geq \log(u_k)/w_k$ to M
    sample $x^{(i)} \sim \mathcal{U}_{SAT(M)}$

5.2   INFINITE DOMAINS

One limitation of Markov logic is that it is only defined for finite domains. While this is seldom a problem in practice, considering the infinite limit can simplify the treatment of some problems, and yield new insights. We would also like to elucidate how far it is possible to combine the full power of first-order logic and graphical models. Thus, in this section we describe how to extend Markov logic to infinite domains, as done by Singla and Domingos [136]. Our treatment is based on the theory of Gibbs measures [38]. Gibbs measures are infinite-dimensional extensions of Markov networks, and have been studied extensively by statistical physicists and mathematical statisticians, due to their importance in modeling systems with phase transitions. We begin with some necessary background on Gibbs measures. We then define MLNs over infinite domains, state sufficient conditions for the existence and uniqueness of a probability measure consistent with a given MLN, and examine the important case of MLNs with non-unique measures. Next, we establish a correspondence between the problem of satisfiability in logic and the existence of MLN measures with certain properties. We conclude with a discussion of the relationship between infinite MLNs and previous infinite relational models.

GIBBS MEASURES

Gibbs measures are infinite-dimensional generalizations of Gibbs distributions. A Gibbs distribution, also known as a log-linear model or exponential model, and equivalent under mild conditions to a Markov network or Markov random field, assigns to a state $\mathbf{x}$ the probability

$$P(\mathbf{X}{=}\mathbf{x}) = \frac{1}{Z} \exp\left(\sum_i w_i f_i(\mathbf{x})\right) \tag{5.1}$$

where w_i is any real number, f_i is an arbitrary function or *feature* of $\mathbf{x}$, and Z is a normalization constant. In this section we will be concerned exclusively with Boolean states and functions (i.e., states are binary vectors, corresponding to possible worlds, and functions are logical formulas). Markov logic can be viewed as the use of first-order logic to compactly specify families of these functions. Thus, a natural way to generalize it to infinite domains is to use the existing theory of Gibbs measures [38]. Although Gibbs measures were primarily developed to model regular lattices (e.g., ferromagnetic materials, gas/liquid phases, etc.), the theory is quite general, and applies equally well to the richer structures definable using Markov logic.

One problem with defining probability distributions over infinite domains is that the probability of most or all worlds will be zero. Measure theory allows us to overcome this problem by instead assigning probabilities to sets of worlds [9]. Let Ω denote the set of all possible worlds, and $\mathcal{E}$ denote a set of subsets of Ω. $\mathcal{E}$ must be a σ-algebra, i.e., it must be non-empty and closed under complements and countable unions. A function $\mu : \mathcal{E} \to \mathbb{R}$ is said to be a *probability measure* over $(\Omega, \mathcal{E})$ if $\mu(E) \geq 0$ for every $E \in \mathcal{E}$, $\mu(\Omega) = 1$, and $\mu(\bigcup E_i) = \sum \mu(E_i)$, where the union is taken over any countable collection of disjoint elements of $\mathcal{E}$.

A related difficulty is that in infinite domains the sum in Equation 5.1 may not exist. However, the distribution of any finite subset of the state variables conditioned on its complement is still well defined. We can thus define the infinite distribution indirectly by means of an infinite collection of finite conditional distributions. This is the basic idea in Gibbs measures.

Let us introduce some notation which will be used throughout the section. Consider a countable set of variables $\mathbf{S} = \{X_1, X_2, \ldots\}$, where each X_i takes values in $\{0, 1\}$. Let $\mathbf{X}$ be a finite set of variables in $\mathbf{S}$, and $\mathbf{S}_\mathbf{X} = \mathbf{S} \setminus \mathbf{X}$. A possible world $\omega \in \Omega$ is an assignment to all the variables in $\mathbf{S}$. Let $\omega_\mathbf{X}$ denote the assignment to the variables in $\mathbf{X}$ under ω, and ω_{X_i} the assignment to X_i. Let $\mathcal{X}$ denote the set of all finite subsets of $\mathbf{S}$. A *basic event* $\mathbf{X} = \mathbf{x}$ is an assignment of values to a finite subset of variables $\mathbf{X} \in \mathcal{X}$, and denotes the set of possible worlds $\omega \in \Omega$ such that $w_\mathbf{X} = \mathbf{x}$. Let $\mathbf{E}$ be the set of all basic events, and let $\mathcal{E}$ be the σ-algebra generated by $\mathbf{E}$, i.e., the smallest σ-algebra containing $\mathbf{E}$. An element E of $\mathcal{E}$ is called an *event*, and $\mathcal{E}$ is the *event space*. The following treatment is adapted from Georgii [38].

Definition 5.2.   An *interaction potential* (or simply a *potential*) is a family $\Phi = (\Phi_\mathbf{V})_{\mathbf{V}\in\mathcal{X}}$ of functions $\Phi_\mathbf{V} : \mathbf{V} \to \mathbb{R}$ such that, for all $\mathbf{X} \in \mathcal{X}$ and $\omega \in \Omega$, the summation

$$H_\mathbf{X}^\Phi(\omega) = \sum_{\mathbf{V}\in\mathcal{X},\mathbf{V}\cap\mathbf{X}\neq\emptyset} \Phi_\mathbf{V}(\omega_\mathbf{V}) \tag{5.2}$$

is finite. $H_\mathbf{X}^\Phi$ is called the Hamiltonian in $\mathbf{X}$ for Φ.

Intuitively, the Hamiltonian $H_{\mathbf{X}}^{\Phi}$ includes a contribution from all the potentials $\Phi_{\mathbf{V}}$ that share at least one variable with the set $\mathbf{X}$. Given an interaction potential Φ and a subset of variables $\mathbf{X}$, we define the conditional distribution $\gamma_{\mathbf{X}}^{\Phi}(\mathbf{X}|\mathbf{S}_{\mathbf{X}})$ as[1]

$$\gamma_{\mathbf{X}}^{\Phi}(\mathbf{X}=\mathbf{x}|\mathbf{S}_{\mathbf{X}}=\mathbf{y}) = \frac{\exp(H_{\mathbf{X}}^{\Phi}(\mathbf{x}, \mathbf{y}))}{\displaystyle\sum_{\mathbf{x}\in\mathrm{Dom}(\mathbf{X})} \exp(H_{\mathbf{X}}^{\Phi}(\mathbf{x}, \mathbf{y}))} \tag{5.3}$$

where the denominator is called the *partition function* in $\mathbf{X}$ for Φ and denoted by $Z_{\mathbf{X}}^{\Phi}$, and $\mathrm{Dom}(\mathbf{X})$ is the domain of $\mathbf{X}$. Equation 5.3 can be easily extended to arbitrary events $E \in \mathcal{E}$ by defining $\gamma_{\mathbf{X}}^{\Phi}(E|\mathbf{S}_{\mathbf{X}})$ to be non-zero only when E is consistent with the assignment in $\mathbf{S}_{\mathbf{X}}$. Details are skipped here to keep the discussion simple, and can be found in Georgii [38]. The family of conditional distributions $\gamma^{\Phi} = (\gamma_{\mathbf{X}}^{\Phi})_{\mathbf{X}\in\mathcal{X}}$ as defined above is called a *Gibbsian specification*.[2]

Given a measure μ over $(\Omega, \mathcal{E})$ and conditional probabilities $\gamma_{\mathbf{X}}^{\Phi}(E|\mathbf{S}_{\mathbf{X}})$, let the composition $\mu\gamma_{\mathbf{X}}^{\Phi}$ be defined as

$$\mu\gamma_{\mathbf{X}}^{\Phi}(E) = \int_{\mathrm{Dom}(\mathbf{S}_{\mathbf{X}})} \gamma_{\mathbf{X}}^{\Phi}(E|\mathbf{S}_{\mathbf{X}})\, \partial\mu \tag{5.4}$$

$\mu\gamma_{\mathbf{X}}^{\Phi}(E)$ is the probability of event E according to the conditional probabilities $\gamma_{\mathbf{X}}^{\Phi}(E|\mathbf{S}_{\mathbf{X}})$ and the measure μ on $\mathbf{S}_{\mathbf{X}}$. We are now ready to define Gibbs measure.

Definition 5.3. Let γ^{Φ} be a Gibbsian specification. Let μ be a probability measure over the measurable space $(\Omega, \mathcal{E})$ such that, for every $\mathbf{X} \in \mathcal{X}$ and $E \in \mathcal{E}$, $\mu(E) = \mu\gamma_{\mathbf{X}}^{\Phi}(E)$. Then the specification γ^{Φ} is said to admit the *Gibbs measure* μ. Further, $\mathcal{G}(\gamma^{\Phi})$ denotes the set of all such measures.

In other words, a Gibbs measure is consistent with a Gibbsian specification if its event probabilities agree with those obtained from the specification. Given a Gibbsian specification, we can ask whether there exists a Gibbs measure consistent with it ($|\mathcal{G}(\gamma^{\Phi})| > 0$), and whether it is unique ($|\mathcal{G}(\gamma^{\Phi})| = 1$). In the non-unique case, we can ask what the structure of $\mathcal{G}(\gamma^{\Phi})$ is, and what the measures in it represent. We can also ask whether Gibbs measures with specific properties exist. The theory of Gibbs measures addresses these questions. In this section, we apply it to the case of Gibbsian specifications defined by MLNs.

DEFINITION

The first-order formulas in an MLN can be converted to equivalent formulas in *prenex conjunctive normal form*, $Qx_1 \ldots Qx_n C(x_1, \ldots, x_n)$, where each Q is a quantifier, the x_i are the quantified variables, and $C(\ldots)$ is a conjunction of clauses. We will assume throughout this section that all existentially quantified variables in an MLN have finite domains, unless otherwise specified. While this is

---

[1] For physical reasons, this equation is usually written with a negative sign in the exponent, i.e., $\exp[-H_{\mathbf{X}}^{\Phi}(\omega)]$. Since this is not relevant in Markov logic and does not affect any of the results, we omit it.

[2] Georgii [38] defines Gibbsian specifications in terms of underlying independent specifications. For simplicity, we assume these to be equi-distributions and omit them throughout this section.

a significant restriction, it still includes essentially all previous probabilistic relational representations as special cases. Existentially quantified formulas can now be replaced by finite disjunctions. By distributing conjunctions over disjunctions, every prenex CNF can now be converted to a quantifier-free CNF, with all variables implicitly universally quantified.

The Herbrand universe $U(L)$ of an MLN L is the set of all ground terms constructible from the constants and function symbols in the MLN. The Herbrand base $B(L)$ of L is the set of all ground atoms and clauses constructible from the predicates in L, the clauses in the CNF form of L, and the terms in $U(L)$, replacing typed variables only by terms of the corresponding type. We assume Herbrand interpretations throughout. We are now ready to formally define infinite MLNs.

Definition 5.4.   A *Markov logic network (MLN)* L is a (finite) set of pairs (F_i, w_i), where F_i is a formula in first-order logic and w_i is a real number. L defines a countable set of variables S and interaction potential $\Phi^L = (\Phi^L_X)_{X \in \mathcal{X}}$, $\mathcal{X}$ being the set of all finite subsets of S, as follows:

1. S contains a binary variable for each atom in $B(L)$. The value of this variable is 1 if the atom is true, and 0 otherwise.

2. $\Phi^L_X(\mathbf{x}) = \sum_j w_j f_j(\mathbf{x})$, where the sum is over the clauses C_j in $B(L)$ whose arguments are exactly the elements of X. If $F_{i(j)}$ is the formula in L from which C_j originated, and $F_{i(j)}$ gave rise to n clauses in the CNF form of L, then $w_j = w_i/n$. $f_j(\mathbf{x}) = 1$ if C_j is true in world $\mathbf{x}$, and $f_j = 0$ otherwise.

For Φ^L to correspond to a well-defined Gibbsian specification, the corresponding Hamiltonians (Equation 5.2) need to be finite. This brings us to the following definition.

Definition 5.5.   Let C be a set of first-order clauses. Given a ground atom $X \in B(C)$, let the *neighbors* $N(X)$ of X be the atoms that appear with it in some ground clause. C is said to be *locally finite* if each atom in the Herbrand base of C has a finite number of neighbors, i.e., $\forall X \in B(C)$, $|N(X)| < \infty$. An MLN (or knowledge base) is said to be locally finite if the set of its clauses is locally finite.

It is easy to see that local finiteness is sufficient to ensure a well-defined Gibbsian specification. Given such an MLN L, the distribution γ^L_X of a set of variables $X \in \mathcal{X}$ conditioned on its complement S_X is given by

$$\gamma^L_X(X=\mathbf{x}|S_X=\mathbf{y}) = \frac{\exp\left(\sum_j w_j f_j(\mathbf{x}, \mathbf{y})\right)}{\sum_{\mathbf{x}' \in \text{Dom}(X)} \exp\left(\sum_j w_j f_j(\mathbf{x}', \mathbf{y})\right)} \tag{5.5}$$

where the sum is over the clauses in $B(L)$ that contain at least one element of X, and $f_j(\mathbf{x}, \mathbf{y}) = 1$ if clause C_j is true under the assignment $(\mathbf{x}, \mathbf{y})$ and 0 otherwise. The corresponding Gibbsian specification is denoted by γ^L.

For an MLN to be locally finite, it suffices that it be σ-*determinate*.

Definition 5.6.   A clause is σ-*determinate* if all the variables with infinite domains it contains appear in all literals.[3] A set of clauses is σ-determinate if each clause in the set is σ-determinate. An MLN is σ-determinate if the set of its clauses is σ-determinate.

Notice that this definition does not require that all literals have the same infinite arguments; for example, the clause $Q(x, y) \Rightarrow R(f(x), g(x, y))$ is σ-determinate. In essence, σ-determinacy requires that the neighbors of an atom be defined by functions of its arguments. Because functions can be composed indefinitely, the network can be infinite; because first-order clauses have finite length, σ-determinacy ensures that neighborhoods are still finite.

If the MLN contains no function symbols, Definition 5.4 reduces to the finite case from Chapter 2 (Definition 2.1), with C being the constants appearing in the MLN. This can be easily seen by substituting $\mathbf{X} = \mathbf{S}$ in Equation 5.5. Notice it would be equally possible to define features for conjunctions of clauses, and this may be preferable for some applications.

EXISTENCE

Let $\mathbf{L}$ be a locally finite MLN. The focus of this subsection is to show that its specification $\gamma^{\mathbf{L}}$ always admits some measure μ. It is useful to first gain some intuition as to why this might not always be the case. Consider an MLN stating that each person is loved by exactly one person: $\forall x \exists^! y \, \text{Loves}(y, x)$. (We the $\exists^!$ quantifier here to mean "exactly one exists.") Let ω_k denote the event $\text{Loves}(P_k, \text{Anna})$, i.e., Anna is loved by the kth person in the (countably infinite) domain. Then, in the limit of infinite weights, one would expect that $\mu(\bigcup \omega_k) = \mu(\Omega) = 1$. But in fact $\mu(\bigcup \omega_k) = \sum \mu(\omega_k) = 0$. The first equality holds because the ω_k's are disjoint, and the second one because each ω_k has zero probability of occurring by itself. There is a contradiction, and there exists no measure consistent with the MLN above.[4] The reason the MLN fails to have a measure is that the formulas are not local, in the sense that the truth value of an atom depends on the truth values of infinite others. Locality is in fact the key property for the existence of a consistent measure, and local finiteness ensures it.

Definition 5.7.   A function $f : \Omega \to \mathbb{R}$ is *local* if it depends only on a finite subset $\mathbf{V} \in \mathcal{X}$. A Gibbsian specification $\gamma = (\gamma_{\mathbf{X}})_{\mathbf{X} \in \mathcal{X}}$ is local if each $\gamma_{\mathbf{X}}$ is local.

Lemma 5.8.   *Let $\mathbf{L}$ be a locally finite MLN, and $\gamma^{\mathbf{L}}$ the corresponding specification. Then $\gamma^{\mathbf{L}}$ is local.*

---

[3]This is related to the notion of a *determinate clause* in logic programming. In a determinate clause, the grounding of the variables in the head determines the grounding of all the variables in the body. In infinite MLNs, any literal in a clause can be inferred from the others, not just the head from the body, so we require that the (infinite-domain) variables in each literal determine the variables in the others.

[4]See Example 4.16 in Georgii [38] for a detailed proof.

Proof. Each Hamiltonian H_X^L is local, since by local finiteness it depends only on a finite number of potentials ϕ_V^L. It follows that each γ_X^L is local, and hence the corresponding specification γ^L is also local. □

We now state the theorem for the existence of a measure admitted by γ^L.

Theorem 5.9. *Let* **L** *be a locally finite MLN, and* $\gamma^L = (\gamma_X^L)_{X \in \mathcal{X}}$ *be the corresponding Gibbsian specification. Then there exists a measure* μ *over* $(\Omega, \mathcal{E})$ *admitted by* γ^L, *i.e.,* $|\mathcal{G}(\gamma^L)| \geq 1$.

Proof. To show the existence of a measure μ, we need to prove the following two conditions:

1. The net $(\gamma_X^L(\mathbf{X}|\mathbf{S_X}))_{X \in \mathcal{X}}$ has a cluster point with respect to the weak topology on $(\Omega, \mathcal{E})$.

2. Each cluster point of $(\gamma_X^L(\mathbf{X}|\mathbf{S_X}))_{X \in \mathcal{X}}$ belongs to $\mathcal{G}(\gamma^L)$.

It is a well known result that, if all the variables X_i have finite domains, then the net in Condition 1 has a cluster point (see Section 4.2 in Georgii [38]). Thus, since all the variables in the MLN are binary, Condition 1 holds. Further, since γ^L is local, every cluster point μ of the net $(\gamma_X^L(\mathbf{X}|\mathbf{S_X}))_{X \in \mathcal{X}}$ belongs to $\mathcal{G}(\gamma^L)$ (Comment 4.18 in Georgii [38]). Therefore, Condition 2 is also satisfied. Hence there exists a measure μ consistent with the specification γ^L, as required. □

UNIQUENESS

We now address the question of under what conditions an MLN admits a unique measure. Let us first gain some intuition as to why an MLN might admit more than one measure. The only condition an MLN **L** imposes on a measure is that it should be consistent with the local conditional distributions γ_X^L. But since these distributions are local, they do not determine the behavior of the measure at infinity. Consider, for example, a semi-infinite two-dimensional lattice, where neighboring sites are more likely to have the same truth value than not. This can be represented by formulas of the form $\forall x, y \; Q(x, y) \Leftrightarrow Q(s(x), y)$ and $\forall x, y \; Q(x, y) \Leftrightarrow Q(x, s(y))$, with a single constant 0 to define the origin $(0, 0)$, and with $s()$ being the successor function. The higher the weight w of these formulas, the more likely neighbors are to have the same value. This MLN has two extreme states: one where $\forall x \; S(x)$, and one where $\forall x \; \neg S(x)$. Let us call these states ξ and $\xi_\neg$, and let ξ' be a local perturbation of ξ (i.e., ξ' differs from ξ on only a finite number of sites). If we draw a contour around the sites where ξ' and ξ differ, then the log odds of ξ and ξ' increase with wd, where d is the length of the contour. Thus long contours are improbable, and there is a measure $\mu \to \delta_\xi$ as $w \to \infty$. Since, by the same reasoning, there is a measure $\mu_\neg \to \delta_{\xi_\neg}$ as $w \to \infty$, the MLN admits more than one measure.[5]

Let us now turn to the mathematical conditions for the existence of a unique measure for a given MLN **L**. Clearly, in the limit of all non-unit clause weights going to zero, **L** defines a unique

---

[5]Notice that this argument fails for a one-dimensional lattice (equivalent to a Markov chain), since in this case an arbitrarily large number of sites can be separated from the rest by a contour of length 2. Non-uniqueness (corresponding to a non-ergodic chain) can then only be obtained by making some weights infinite (corresponding to zero transition probabilities).

distribution. Thus, by a continuity argument, one would expect the same to be true for small enough weights. This is indeed the case. To make it precise, let us first define the notion of the oscillation of a function. Given a function $f : \mathbf{X} \to \mathbb{R}$, let the oscillation of f, $\delta(f)$, be defined as

$$
\begin{aligned}
\delta(f) &= \max_{\mathbf{x},\mathbf{x}' \in \mathrm{Dom}(\mathbf{X})} |f(\mathbf{x}) - f(\mathbf{x}')| \\
&= \max_{\mathbf{x}} |f(\mathbf{x})| - \min_{\mathbf{x}} |f(\mathbf{x})| \tag{5.6}
\end{aligned}
$$

The oscillation of a function is thus simply the difference between its extreme values. We can now state a sufficient condition for the existence of a unique measure.

Theorem 5.10. *Let $\mathbf{L}$ be a locally finite MLN with interaction potential Φ^L and Gibbsian specification γ^L such that*

$$
\sup_{X_i \in S} \sum_{C_j \in \mathbf{C}(X_i)} (|C_j| - 1)|w_j| < 2 \tag{5.7}
$$

where $\mathbf{C}(X_i)$ is the set of ground clauses in which X_i appears, $|C_j|$ is the number of ground atoms appearing in clause C_j, and w_j is its weight. Then γ^L admits a unique Gibbs measure.

Proof. Based on Theorem 8.7 and Proposition 8.8 in Georgii [38], a sufficient condition for uniqueness is

$$
\sup_{X_i \in S} \sum_{\mathbf{V} \ni X_i} (|\mathbf{V}| - 1)\delta(\Phi^\mathbf{L}_\mathbf{V}) < 2 \tag{5.8}
$$

Rewriting this condition in terms of the ground formulas in which a variable X_i appears (see Definition 5.4) yields the desired result. □

Note that, as alluded to before, the above condition does not depend on the weight of the unit clauses. This is because for a unit clause $|C_j| - 1 = 0$. If we define the interaction between two variables as the sum of the weights of all the ground clauses in which they appear together, then the above theorem states that the total sum of the interactions of any variable with its neighbors should be less than 2 for the measure to be unique.

Two other sufficient conditions are worth mentioning briefly. One is that, if the weights of the unit clauses are sufficiently large compared to the weights of the non-unit ones, the measure is unique. Intuitively, the unit terms "drown out" the interactions, rendering the variables approximately independent. The other condition is that, if the MLN is a one-dimensional lattice, it suffices that the total interaction between the variables to the left and right of any edge be finite. This corresponds to the ergodicity condition for a Markov chain.

NON-UNIQUE MLNS

At first sight, it might appear that non-uniqueness is an undesirable property, and non-unique MLNs are not an interesting object of study. However, the non-unique case is in fact quite important, because many phenomena of interest are represented by MLNs with non-unique measures (for example, very large social networks with strong word-of-mouth effects). The question of what these measures represent, and how they relate to each other, then becomes important. This is the subject of this subsection.

The first observation is that the set of all Gibbs measures $\mathcal{G}(\gamma^L)$ is convex. That is, if $\mu, \mu' \in \mathcal{G}(\gamma^L)$ then $\nu \in \mathcal{G}(\gamma^L)$, where $\nu = s\mu + (1 - s)\mu', s \in (0, 1)$. This is easily verified by substituting ν in Equation 5.4. Hence, the non-uniqueness of a Gibbs measure implies the existence of infinitely many consistent Gibbs measures. Further, many properties of the set $\mathcal{G}(\gamma^L)$ depend on the set of extreme Gibbs measures ex $\mathcal{G}(\gamma^L)$, where $\mu \in$ ex $\mathcal{G}(\gamma^L)$ if $\mu \in \mathcal{G}(\gamma^L)$ cannot be written as a linear combination of two distinct measures in $\mathcal{G}(\gamma^L)$.

An important notion to understand the properties of extreme Gibbs measures is the notion of a tail event. Consider a subset $\mathbf{S}'$ of $\mathbf{S}$. Let $\sigma(\mathbf{S}')$ denote the σ-algebra generated by the set of basic events involving only variables in $\mathbf{S}'$. Then we define the tail σ-algebra $\mathcal{T}$ as

$$\mathcal{T} = \bigcap_{X \in \mathcal{X}} \sigma(\mathbf{S}_X) \tag{5.9}$$

Any event belonging to $\mathcal{T}$ is called a tail event. $\mathcal{T}$ is precisely the set of events which do not depend on the value of any finite set of variables, but rather only on the behavior at infinity. For example, in the infinite tosses of a coin, the event that ten consecutive heads come out infinitely many times is a tail event. Similarly, in the lattice example in the previous section, the event that a finite number of variables have the value 1 is a tail event. Events in $\mathcal{T}$ can be thought of as representing macroscopic properties of the system being modeled.

Definition 5.11. A measure μ is *trivial* on a σ-algebra $\mathcal{E}$ if $\mu(E) = 0$ or 1 for all $E \in \mathcal{E}$.

The following theorem (adapted from Theorem 7.8 in Georgii [38]) describes the relationship between the extreme Gibbs measures and the tail σ-algebra.

Theorem 5.12. *Let* $\mathbf{L}$ *be a locally finite MLN, and* γ^L *denote the corresponding Gibbsian specification. Then the following results hold:*

1. *A measure* $\mu \in$ *ex* $\mathcal{G}(\gamma^L))$ *iff it is trivial on the tail* σ*-algebra* $\mathcal{T}$.

2. *Each measure* μ *is uniquely determined by its behavior on the tail* σ*-algebra, i.e., if* $\mu_1 = \mu_2$ *on* $\mathcal{T}$ *then* $\mu_1 = \mu_2$.

It is easy to see that each extreme measure corresponds to some particular value for all the macroscopic properties of the network. In physical systems, extreme measures correspond to phases of the system (e.g., liquid vs. gas, or different directions of magnetization), and non-extreme measures correspond to probability distributions over phases. Uncertainty over phases arises when our knowledge of a system is not sufficient to determine its macroscopic state. Clearly, the study of non-unique MLNs beyond the highly regular ones statistical physicists have focused on promises to be quite interesting. In the next subsection, we take a step in this direction by considering the problem of satisfiability in the context of MLN measures.

SATISFIABILITY AND ENTAILMENT

From Theorem 2.6, we know that MLNs generalize finite first-order logic in the infinite-weight limit. We now extend this result to infinite domains.

Consider an MLN $\mathbf{L}$ such that each clause in its CNF form has the same weight w. In the limit $w \to \infty$, $\mathbf{L}$ does not correspond to a valid Gibbsian specification since the Hamiltonians defined in Equation 5.2 are no longer finite. Revisiting Equation 5.5 in the limit of all equal infinite clause weights, the limiting conditional distribution is equi-distribution over those configurations $\mathbf{X}$ that satisfy the maximum number of clauses given $\mathbf{S_X} = \mathbf{y}$. It turns out we can still talk about the existence of a measure consistent with these conditional distributions because they constitute a valid specification (though not Gibbsian) under the same conditions as in the finite weight case. Details and proofs can be found in Singla and Domingos [137]. Existence of a measure follows as in the case of finite weights because of the locality of conditional distributions. We now define the notion of a *satisfying measure*, which is central to the results presented in this section.

Definition 5.13. Let $\mathbf{L}$ be a locally finite MLN. Given a clause $C_i \in \mathbf{B}(\mathbf{L})$, let $\mathbf{V}_i$ denote the set of Boolean variables appearing in C_i. A measure $\mu \in \mathcal{G}(\gamma^{\mathbf{L}})$ is said to be a *satisfying measure* for $\mathbf{L}$ if, for every ground clause $C_i \in \mathbf{B}(\mathbf{L})$, μ assigns non-zero probability only to the satisfying assignments of the variables in C_i, i.e., $\mu(\mathbf{V}_i = \mathbf{v}_i) > 0$ implies that $\mathbf{V}_i = \mathbf{v}_i$ is a satisfying assignment for C_i. $\mathcal{S}(\gamma^{\mathbf{L}})$ denotes the set of all satisfying measures for $\mathbf{L}$.

Informally, a satisfying measure assigns non-zero probability only to those worlds that are consistent with all the formulas in $\mathbf{L}$. Intuitively, existence of a satisfying measure for $\mathbf{L}$ should be in some way related to the existence of a satisfying assignment for the corresponding knowledge base. Our next theorem formalizes this intuition.

Theorem 5.14. *Let $\mathbf{K}$ be a locally finite knowledge base, and let $\mathbf{L}_\infty$ be the MLN obtained by assigning weight $w \to \infty$ to all the clauses in $\mathbf{K}$. Then there exists a satisfying measure for $\mathbf{L}_\infty$ iff $\mathbf{K}$ is satisfiable. Mathematically,*

$$|\mathcal{S}(\gamma^{\mathbf{L}_\infty})| > 0 \iff \textit{Satisfiable}(\mathbf{K}) \tag{5.10}$$

A full proof can be found in Singla and Domingos [137].

Corollary 5.15. *Let* **K** *be a locally finite knowledge base. Let* α *be a first-order formula, and* $\mathbf{L}_\infty^\alpha$ *be the MLN obtained by assigning weight* $w \to \infty$ *to all clauses in* $\mathbf{K} \cup \{\neg\alpha\}$. *Then* **K** *entails* α *iff* $\mathbf{L}_\infty^\alpha$ *has no satisfying measure. Mathematically,*

$$\mathbf{K} \models \alpha \;\Leftrightarrow\; |\mathcal{S}(\gamma^{\mathbf{L}_\infty^\alpha})| = 0 \tag{5.11}$$

Thus, for locally finite knowledge bases with Herbrand interpretations, first-order logic can be viewed as the limiting case of Markov logic when all weights tend to infinity. Whether these conditions can be relaxed is a question for future work.

5.3 RECURSIVE MARKOV LOGIC

In Markov logic, the unification of logic and probability is incomplete. Markov logic only treats the top-level conjunction and universal quantifiers in a knowledge base as probabilistic, when in principle any logical combination can be viewed as the limiting case of an underlying probability distribution; disjunctions and existential quantifiers remain deterministic. Thus the symmetry between conjunctions and disjunctions, and between universal and existential quantifiers, is lost (except in the infinite-weight limit).

For example, an MLN with the formula $R(X) \wedge S(X)$ can treat worlds that violate both $R(X)$ and $S(X)$ as less probable than worlds that only violate one. Since an MLN acts as a soft conjunction, the groundings of $R(X)$ and $S(X)$ simply appear as distinct formulas. (As usual, we will assume that the knowledge base is converted to CNF before performing learning or inference.) This is not possible for the disjunction $R(X) \vee S(X)$: no distinction is made between satisfying both $R(X)$ and $S(X)$ and satisfying just one. Since a universally quantified formula is effectively a conjunction over all its groundings, while an existentially quantified formula is a disjunction over them, this leads to the two quantifiers being handled differently.

This asymmetry can be avoided by "softening" disjunction and existential quantification in the same way that Markov logic softens conjunction and universal quantification. The result is a representation in which MLNs can have nested MLNs as features. We call these recursive Markov logic networks, or *recursive random fields (RRFs)* for short.

RRFs have many desirable properties, including the ability to represent distributions like noisy DNF, rules with exceptions, and *m*-of-all quantifiers much more compactly than MLNs. RRFs also allow more flexibility in revising first-order theories to maximize data likelihood. Standard methods for inference in Markov networks are easily extended to RRFs, and weight learning can be carried out efficiently using a variant of the backpropagation algorithm.

RRF theory revision can be viewed as a first-order probabilistic analog of the KBANN algorithm, which initializes a neural network with a propositional theory and uses backpropagation to improve its fit to data [145]. A propositional RRF (where all predicates have zero arity) differs from a multilayer perceptron in that its output is the joint probability of its inputs, not the regression of a variable on others (or, in the probabilistic version, its conditional probability). Propositional RRFs are an alternative to Boltzmann machines, with nested features playing the role of hidden variables. Because the nested features are deterministic functions of the inputs, learning does not require EM, and inference does not require marginalizing out variables.

A recursive random field is a log-linear model in which each feature is either an observable random variable or the output of another recursive random field. To build up intuition, we first describe the propositional case, then generalize it to the more interesting relational case. A concrete example is given in a later subsection, and illustrated in Figure 5.1.

PROPOSITIONAL RRFS

While our primary goal is solving relational problems, RRFs may be interesting in propositional domains as well. Propositional RRFs extend Markov random fields and Boltzmann machines in the same way multilayer perceptrons extend single-layer ones. The extension is very simple in principle, but allows RRFs to compactly represent important concepts, such as m-of-n. It also allows RRFs to learn features via weight learning, which could be more effective than current feature-search methods for Markov random fields.

The probability distribution represented by a propositional RRF is as follows:

$$P(\mathbf{X} = \mathbf{x}) = \frac{1}{Z_0} \exp \left(\sum_i w_i f_i(\mathbf{x}) \right)$$

where Z_0 is a normalization constant, to ensure that the probabilities of all possible states $\mathbf{x}$ sum to 1. What makes this different from a standard Markov network is that the features can be built up from other subfeatures to an arbitrary number of levels. Specifically, each $f_i(\mathbf{x})$ is either:

$$f_i(\mathbf{x}) = x_j \quad \text{(base case), or}$$
$$f_i(\mathbf{x}) = \frac{1}{Z_i} \exp \left(\sum_j w_{ij} f_j(\mathbf{x}) \right) \quad \text{(recursive case)}$$

In the recursive case, the summation is over all features f_j referenced by the "parent" feature f_i. A child feature, f_j, can appear in more than one parent feature, and thus an RRF can be viewed as a directed acyclic graph of features. The attribute values are at the leaves, and the probability of their configuration is given by the root. (Note that the probabilistic graphical model represented by the RRF is still undirected.)

Since the overall distribution is simply a recursive feature, we can also write the probability distribution as follows:

$$P(\mathbf{X} = \mathbf{x}) = f_0(\mathbf{x})$$

Except for Z_0 (the normalization of the root feature, f_0), the per-feature normalization constants Z_i can be absorbed into the corresponding feature weights w_{ki} in their parent features f_k. Therefore, the user is free to choose any convenient normalization, or even no normalization at all.

It is easy to show that this generalizes Markov networks with conjunctive or disjunctive features. Each f_i approximates a conjunction when weights w_{ij} are very large. In the limit, f_i will be 1 iff the conjunct is true. f_i can also represent disjunctions using large negative weights, along with a negative weight for the parent feature f_k, w_{ki}. The negative weight w_{ki} turns the conjunction into a disjunction just as negation does in De Morgan's laws. However, one can also move beyond conjunction and disjunction to represent m-of-n concepts, or even more complex distributions where different features have different weights.

Note that features with small absolute weights have little effect. Therefore, instead of using heuristics or search to determine which attributes should appear in which feature, we can include *all* predicates and let weight learning sort out which attributes are relevant for which feature. This is similar to learning a neural network by initializing it with small random values. Since the network can represent any logical formula, there is no need to commit to a specific structure ahead of time. This is an attractive alternative to the traditional inductive methods used for learning MRF features.

An RRF can be seen as a type of multi-layer neural network, in which the node function is exponential (rather than sigmoidal) and the network is trained to maximize joint likelihood. Unlike in multilayer perceptrons, where some random variables are inputs and some are outputs, in RRFs all variables are inputs, and the output is their joint probability. In other ways, an RRF resembles a Boltzmann machine, but with the greater flexibility of multiple layers and learnable using a variant of the back-propagation algorithm. RRFs have no hidden variables to sum out since all nodes in the network have deterministic values, making inference more efficient.

RELATIONAL RRFS

In the relational case, relations over an arbitrary number of objects take the place of a fixed number of variables. To allow parameter tying across different groundings, we use parameterized features, or *parfeatures*. We represent the parameter tuple as a vector, $\vec{g}$, whose size depends on the arity of the parfeature. Note that $\vec{g}$ is a vector of logical variables (i.e., arguments to predicates) as opposed to the random Boolean variables $\mathbf{x}$ (ground atoms) that represent a state of the world. We use subscripts to distinguish among parfeatures with different parameterizations, e.g. $f_{i,\vec{g}}(\mathbf{x})$ and $f_{i,\vec{g}'}(\mathbf{x})$ represent different groundings of the ith parfeature.

Each RRF parfeature is defined in one of two ways:

$$f_{i,\vec{g}}(\mathbf{x}) = \mathrm{R_i}(\mathrm{g_{i_1}}, \ldots, \mathrm{g_{i_k}}) \quad \text{(base case)}$$

$$f_{i,\vec{g}}(\mathbf{x}) = \frac{1}{Z_i} \exp\left(\sum_j w_{ij} \sum_{\vec{g}'} f_{j,\vec{g},\vec{g}'}(\mathbf{x})\right) \quad \text{(recursive case)}$$

The base case is straightforward: it simply represents the truth value of a ground relation (as specified by $\mathbf{x}$). There is one such grounding for each possible combination of parameters (arguments) of the parfeature. The recursive case sums the weighted values of all child parfeatures. Each parameter g_i of a child parfeature is either a parameter of the parent feature ($g_i \in \vec{g}$) or a parameter of a child feature that is summed out and does not appear in the parent feature ($g_i \in \vec{g}'$). (These $\vec{g}'$ parameters are analogous to the parameters that appear in the body but not the head of a Horn clause.) Just as sums of child features act as conjunctions, the summations over $\vec{g}'$ parameters act as universal quantifiers with Markov logic semantics. In fact, these generalized quantifiers can represent m-of-all concepts, just as the simple feature sums can represent m-of-n concepts.

The relational version of a recursive random field is therefore defined as follows:

$$P(\mathbf{X} = \mathbf{x}) = f_0(\mathbf{x})$$

where $\mathbf{X}$ is the set of all ground relations (e.g., R(A, B), S(A)), $\mathbf{x}$ is an assignment of truth values to ground relations, and f_0 is the root recursive parfeature (which, being the root, has no parameters). Since f_0 is a recursive parfeature, it is normalized by the constant Z_0 to ensure a valid probability distribution. (As in the propositional case, all other Z_i's can be absorbed into the weights of their parent features, and may therefore be normalized in any convenient way.)

Any relational RRF can be converted into a propositional RRF by grounding all parfeatures and expanding all summations. Each distinct grounding of a parfeature becomes a distinct feature, but with shared weights.

RRF EXAMPLE

To clarify these ideas, let us take the example knowledge base from Chapter 2. The domain consists of three predicates: Smokes(g) (g is a smoker); Cancer(g) (g has cancer); and Friends(g, h) (g is a friend of h). We abbreviate these predicates as Sm(g), Ca(g), and Fr(g, h), respectively.

We wish to represent three beliefs: (i) smoking causes cancer; (ii) friends of friends are friends (transitivity of friendship); and (iii) everyone has at least one friend who smokes. (The most interesting belief from Chapter 2, that people tend to smoke if their friends do, is omitted here for simplicity.) We demonstrate how to represent these beliefs by first converting them to first-order logic, and then converting to an RRF.

One can represent the first belief, "smoking causes cancer," in first-order logic as a universally quantified implication: $\forall g\ \text{Sm}(g) \Rightarrow \text{Ca}(g)$. This implication can be rewritten as a disjunction: $\neg\text{Sm}(g) \vee \text{Ca}(g)$. From De Morgan's laws, this is equivalent to: $\neg(\text{Sm}(g) \wedge \neg\text{Ca}(g))$, which can be represented as an RRF feature:

$$f_{1,g}(\mathbf{x}) = \frac{1}{Z_1} \exp(w_{1,1}\text{Sm}(g) + w_{1,2}\text{Ca}(g))$$

where $w_{1,1}$ is positive, $w_{1,2}$ is negative, and the feature weight $w_{0,1}$ is negative (not shown above). In general, since RRF features can model conjunction and disjunction, any CNF knowledge base can be represented as an RRF. A similar approach works for the second belief, "friends of people are friends."

The first two beliefs are also handled well by Markov logic networks. The key advantage of recursive random fields is in representing more complex formulas. The third belief, "everyone has at least one friend who smokes," is naturally represented by nested quantifiers: $\forall g \exists h \, \text{Fr}(g, h) \wedge \text{Sm}(h)$. This is best represented as an RRF feature that references a secondary feature:

$$f_{3,g}(\mathbf{x}) = \frac{1}{Z_3} \exp \left(\sum_h w_{3,1} f_{4,g,h}(\mathbf{x}) \right)$$

$$f_{4,g,h}(\mathbf{x}) = \frac{1}{Z_4} \exp(w_{4,1} \text{Fr}(g, h) + w_{4,2} \text{Sm}(h))$$

Note that in RRFs this feature can also represent a distribution over the number of smoking friends each person has, depending on the assigned weights. It is possible that, while almost everyone has at least one smoking friend, many people have at least two or three. With an RRF, we can actually learn this distribution from data.

This third belief is very problematic for an MLN. First of all, in an MLN it is purely logical: there is no change in probability with the number of smoking friends once that number exceeds one. Secondly, MLNs do not represent the belief efficiently. In an MLN, the existential quantifier is converted to a very large disjunction:

$$(\text{Fr}(g, A) \wedge \text{Sm}(A)) \vee (\text{Fr}(g, B) \wedge \text{Sm}(B)) \vee \cdots$$

If there are 1000 objects in the database, then this disjunction is over 1000 conjunctions. Further, converting this to clausal form would require 2^{1000} CNF clauses for each grounding of this rule.

These features define a full joint distribution as follows:

$$P(\mathbf{X} = \mathbf{x}) = \frac{1}{Z_0} \exp \left(\sum_g w_{0,1} f_{1,g}(\mathbf{x}) + \sum_{g,h,i} w_{0,2} f_{2,g,h,i}(\mathbf{x}) + \sum_g w_{0,3} f_{3,g}(\mathbf{x}) \right)$$

Figure 5.1 diagrams the first-order knowledge base containing all of these beliefs, along with the corresponding RRF.

INFERENCE

Of the inference methods discussed in Chapter 3, MaxWalkSAT and MC-SAT are not directly applicable because they exploit the logical structure of MLNs, which RRFs do not share. Instead, Gibbs sampling is used for computing probabilities and iterated conditional modes (ICM) [7] is used for approximating the most likely state. ICM starts from a random configuration and sets each variable in turn to its most likely value, conditioned on all other variables. This procedure continues until no single-variable change will further improve the probability. ICM is easy to implement, fast to run, and guaranteed to converge. Unfortunately, it has no guarantee of converging to the most likely overall configuration. Possible improvements include random restarts, simulated annealing, etc. ICM can also be used to find a good initial state for the Gibbs sampling, significantly reducing burn-in time and achieving better predictions sooner.

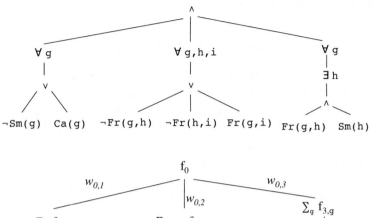

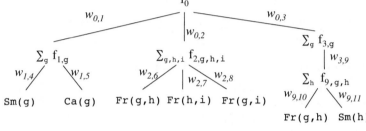

Figure 5.1: Comparison of first-order logic and RRF structures. The RRF structure closely mirrors that of first-order logic, but connectives and quantifiers are replaced by weighted sums.

LEARNING

Given a particular RRF structure and initial set of weights, we can learn weights using a novel variant of the back-propagation algorithm. As in traditional back-propagation, the goal is to efficiently compute the derivative of the loss function with respect to each weight in the model. In this case, the loss function is not the error in predicting the output variables, but rather the joint log likelihood of all variables. We must also consider the partition function for the root feature, Z_0. For these computations, we extract the $1/Z_0$ term from f_0, and use f_0 refer to the unnormalized feature value.

We begin by discussing the simpler, propositional case. We abbreviate $f_i(\mathbf{x})$ as f_i for these arguments. The derivative of the log likelihood with respect to a weight w_{ij} consists of two terms:

$$\frac{\partial \log P(\mathbf{x})}{\partial w_{ij}} = \frac{\partial \log(f_0/Z_0)}{\partial w_{ij}} = \frac{\partial \log(f_0)}{\partial w_{ij}} - \frac{\partial \log(Z_0)}{\partial w_{ij}}$$

The first term can be evaluated with the chain rule:

$$\frac{\partial \log(f_0)}{\partial w_{ij}} = \frac{\partial \log(f_0)}{\partial f_i} \frac{\partial f_i}{\partial w_{ij}}$$

From the definition of f_i (including the normalization Z_i):

$$\frac{\partial f_i}{\partial w_{ij}} = f_i \left(f_j - \frac{1}{Z_i} \frac{\partial Z_i}{\partial w_{ij}} \right)$$

From repeated applications of the chain rule, the $\partial \log(f_0)/\partial f_i$ term is the sum of all derivatives along all paths through the network from f_0 to f_i. Given a path in the feature graph $\{ f_0, f_a, f_b, \ldots, f_k, f_i \}$, the derivative along that path takes the form $f_0 w_a f_a w_b f_b \cdots w_k f_k w_i$. We can efficiently compute the sum of all paths by caching the per-feature partials, $\partial f_0/\partial f_i$, analogous to back-propagation.

The second term, $\partial \log(Z_0)/\partial w_{ij}$, is the expected value of the first term, evaluated over all possible inputs $\mathbf{x}'$. Therefore, the complete partial derivative is:

$$\frac{\partial \log P(\mathbf{x})}{\partial w_{ij}} = \frac{\partial \log(f_0(\mathbf{x}))}{\partial w_{ij}} - E_{\mathbf{x}'} \left[\frac{\partial \log(f_0(\mathbf{x}'))}{\partial w_{ij}} \right]$$

where the individual components are evaluated as above.

Computing the expectation is typically intractable, but it can be approximated using Gibbs sampling. A more efficient alternative is to instead optimize the pseudo-likelihood [6], as done in generative training of MLNs (Section 4.1).

The expression for the gradient of the pseudo-log-likelihood of a propositional RRF is as follows:

$$\frac{\partial \log P^*(\mathbf{X}=\mathbf{x})}{\partial w_i} = \sum_{t=1}^{n} P(X_t = \neg x_t | MB_x(X_t)) \times \left(\frac{\partial \log f_0}{\partial w_i} - \frac{\partial \log f_{0[X_t=\neg x_t]}}{\partial w_i} \right)$$

We can compute this by iterating over all query predicates, toggling each one in turn, and computing the relative likelihood and unnormalized likelihood gradient for that permuted state. Note that we compute the gradient of the unnormalized log likelihood as a subroutine in computing the gradient of the pseudo-log-likelihood. However, we no longer need to approximate the intractable normalization term, Z.

To learn a relational RRF, we use the domain to instantiate a propositional RRF with tied weights. The number of features as well as the number of children per feature will depend on the number of objects in the domain. Instead of a weight being attached to a single feature, it is now attached to a set of groundings of a parfeature. The partial derivative with respect to a weight is therefore the sum of the partial derivatives with respect to each instantiation of the shared weight.

RRFS VS. MLNS

Both RRFs and MLNs subsume probabilistic models and first-order logic in finite domains. Both can be trained generatively or discriminatively using gradient descent, either to optimize log likelihood or pseudo-likelihood. For both, when optimizing log likelihood, the normalization constant Z_0 can be approximated using the MPE state or MCMC.

Any MLN can be converted into a relational RRF by translating each clause into an equivalent parfeature. With sufficiently large weights, a parfeature approximates a hard conjunction or disjunction over its children. However, when its weights are sufficiently distinct, a parfeature can take on a different value for each configuration of its children. This allows RRFs to compactly represent distributions that would require an exponential number of clauses in an MLN.

Any RRF can be converted to an MLN by flattening the model, but this will typically require an exponential number of clauses. Such an MLN would be intractable for learning or inference. RRFs are therefore much better at modeling soft disjunction, existential quantification, and nested formulas.

In addition to being "softer" than an MLN clause, an RRF parfeature can represent many different MLN clauses simply by adjusting its weights. This makes RRF weight learning more powerful than MLN structure learning: an RRF with $n + 1$ recursive parfeatures (one for the root) can represent any MLN structure with up to n clauses, as well as many distributions that an n-clause MLN cannot represent.

This leads to new alternatives for structure learning and theory revision. In a domain where little background knowledge is available, an RRF could be initialized with small random weights and still converge to a good statistical model. This is potentially much better than MLN structure learning, which constructs clauses one predicate at a time, and must adjust weights to evaluate every candidate clause.

When background knowledge is available, we can begin by initializing the RRF to the background theory, just as in MLNs. However, in addition to the known dependencies, we can also add dependencies on other parfeatures or predicates with very small weights. Weight learning can learn large weights for relevant dependencies and negligible weights for irrelevant dependencies. This is analogous to what the KBANN system does using neural networks [145]. In contrast, MLNs can only do theory revision through discrete search.

With the current algorithms, MLNs remain much more efficient than RRFs. This is partly because more work has been done on MLN learning and inference, including optimized implementations in Alchemy, and partly due to Markov logic's more restricted structure. RRFs are also very sensitive to initial conditions during weight optimization. Due to these factors, MLNs remain a more practical solution for most applications. Coming up with more efficient RRF algorithms and implementations is an important area for future work. RRFs are not currently available in Alchemy.

5.4   RELATIONAL DECISION THEORY

The problem of extending decision theory [4] to statistical relational models had remained largely unaddressed to date. The one major exception is relational reinforcement learning and first-order MDPs. (e.g., [32, 146, 127]). However, the representations and algorithms in these approaches are geared to the problem of sequential decision-making, and many decision-theoretic problems are not sequential. In particular, relational domains often lead to very large and complex decision problems for which no effective general solution is currently available (e.g., influence maximization in social networks, combinatorial auctions with uncertain supplies, etc.).

In this section, we describe a framework introduced by Nath and Domingos [89] for relational decision theory using Markov logic. This framework allows for very rich utility functions by attaching utility weights as well as probability weights to clauses. In particular, both classical planning and Markov decision processes are special cases of this framework. Maximizing expected utility in this representation is intractable, but there are approximate algorithms for it, based on a combination of weighted satisfiability testing for maximization and MCMC for computing expectations.

MARKOV DECISION NETWORKS

An *influence diagram* or *decision network* is a graphical representation of a decision problem [48]. It consists of a Bayesian network augmented with two types of nodes: *decision* or *action* nodes and *utility* nodes. The action nodes represent the agent's choices; factors involving these nodes and *state* nodes in the Bayesian network represent the (probabilistic) effect of the actions on the world. *Utility* nodes represent the agent's utility function, and are connected to the state nodes that directly influence utility. We can also define a *Markov decision network* as a decision network with a Markov network instead of a Bayesian network.

The fundamental inference problem in decision networks is finding the assignment of values to the action nodes that maximizes the agent's expected utility, possibly conditioned on some evidence. If $\mathbf{a}$ is a choice of actions, $\mathbf{e}$ is the evidence, $\mathbf{x}$ is a state, and $U(\mathbf{x}|\mathbf{a}, \mathbf{e})$ is the utility of $\mathbf{x}$ given $\mathbf{a}$ and $\mathbf{e}$, then the *MEU problem* is to compute $\operatorname{argmax}_{\mathbf{a}} E[U(\mathbf{x}|\mathbf{a}, \mathbf{e})] = \operatorname{argmax}_{\mathbf{a}} \sum_{\mathbf{x}} P(\mathbf{x}|\mathbf{a}, \mathbf{e})U(\mathbf{x}|\mathbf{a}, \mathbf{e})$.

MARKOV LOGIC DECISION NETWORKS

Decision theory can be incorporated into Markov logic simply by allowing formulas to have utilities as well as weights. This puts the expressiveness of first-order logic at our disposal for defining utility functions, at the cost of very little additional complexity in the language. Let an *action predicate* be a predicate whose groundings correspond to possible actions (choices, decisions) by the agent, and a *state predicate* be any predicate in a standard MLN. Formally:

Definition 5.16. A *Markov logic decision network (MLDN)* L is a set of triples (F_i, w_i, u_i), where F_i is a formula in first-order logic and w_i and u_i are real numbers. Together with a finite set of constants $C = \{c_1, c_2, \ldots, c_{|C|}\}$, it defines a Markov decision network $M_{L,C}$ as follows:

1. $M_{L,C}$ contains one binary node for each possible grounding of each state and action predicate appearing in L. The value of the node is 1 if the ground atom is true, and 0 otherwise.

2. $M_{L,C}$ contains one feature for each possible grounding of each formula F_i in L for which $w_i \neq 0$. The value of this feature is 1 if the ground formula is true, and 0 otherwise. The weight of the feature is the w_i associated with F_i in L.

3. $M_{L,C}$ contains one utility node for each possible grounding of each formula F_i in L for which $u_i \neq 0$. The value of the node is the utility u_i associated with F_i in L if F_i is true, and 0 otherwise.

We refer to groundings of action predicates as *action atoms*, and groundings of state predicates as *state atoms*. An assignment of truth values to all action atoms is an *action choice*. An assignment of truth values to all state atoms is a *state of the world* or *possible world*. The utility of world $\mathbf{x}$ given action choice $\mathbf{a}$ and evidence $\mathbf{e}$ is $U(\mathbf{x}|\mathbf{a}, \mathbf{e}) = \sum_i u_i n_i(\mathbf{x}, \mathbf{a}, \mathbf{e})$, where n_i is the number of true groundings of F_i. The expected utility of action choice $\mathbf{a}$ given evidence $\mathbf{e}$ is:

$$E[U(\mathbf{x}|\mathbf{a}, \mathbf{e})] = \sum_{\mathbf{x}} P(\mathbf{x}|\mathbf{a}, \mathbf{e}) \sum_i u_i n_i(\mathbf{x}, \mathbf{a}, \mathbf{e}) = \sum_i u_i E[n_i] \tag{5.12}$$

This expectation can be computed using existing algorithms such as MC-SAT.

A wide range of decision problems can be elegantly formulated as MLDNs, including both classical planning and Markov decision processes (MDPs). To represent an MDP as an MLDN, we can define a constant for each state, action and time step, and the predicates State(s!, t) and Action(a!, t), with the obvious meaning. (The ! notation indicates that, for each t, exactly one grounding of State(s, t) is true.) The transition function is then represented by the formula State(+s, t) ∧ Action(+a, t) ⇒ State(+s', t + 1), with a separate weight for each (s, a, s') triple. (Formulas with + signs before certain variables represent sets of identical formulas with separate weights, one for each combination of groundings of the variables with + signs.) The reward function is defined by the unit clause State(∗s, t), with a utility for each state (using ∗ to represent per-grounding utilities). Policies can be represented by formulas of the form State(+s, t) ⇒ Action(+a, t). Infinite-horizon MDPs can be represented using infinite MLNs (Section 5.2). Partially-observable MDPs are represented by adding the observation model: State(+s, t) ⇒ Observation(+o, t).

Since classical planning languages are variants of first-order logic, translating problems formulated in these languages into MLDNs is straightforward. For simplicity, suppose the problem has been expressed in satisfiability form [52]. It suffices then to translate the (first-order) CNF into a deterministic MLN by assigning infinite weight to all clauses, and to assign a positive utility to the formula defining the goal states. MLDNs now offer a path to extend classical planning with uncertain actions, complex utilities, etc., by assigning finite weights and utilities to formulas. (For example, an action with uncertain effects can be represented by assigning a finite weight to the axiom that defines them.) This can be used to represent first-order MDPs in a manner analogous to Boitilier *et al.* [10].

MAXIMIZING EXPECTED UTILITY

The problem of selecting an action grounding assignment that maximizes expected utility has parallels with the problem of selecting a truth assignment that maximizes the probability of the world. Table 5.3 gives a modified version of the MaxWalkSAT algorithm that maximizes expected utility instead of probability. We refer to this algorithm as RMEU (Relational Maximization of Expected Utility).

Table 5.3: Algorithm for maximizing expected utility in an MLDN.

---

function RMEU(L, DB, n, m_t, m_f)
  inputs: L, a Markov logic decision network
  DB, database containing evidence
  n, number of restarts
  m_t, number of tries
  m_f, number of flips
  output: highest utility configuration of the action atoms found

for $i \leftarrow 1$ **to** n
  Run WalkSAT to satisfy all hard clauses.
  for $j \leftarrow 1$ **to** m_t
    for $k \leftarrow 1$ **to** m_f
      $action \leftarrow$ Random action ground atom
      Flip $action$.
      Run WalkSAT to satisfy hard clauses.
      if flip violates hard clauses
        Flip $action$ back.
      else
        Infer probabilities of ground clauses.
        $utility \leftarrow \sum_{i=1}^{F} u_i \sum_{j=1}^{g_i} p_{ij}$
return solution with highest utility found.

Computing Probabilities

In each iteration of RMEU, we need to infer the probabilities of all ground clauses with non-zero utilities. A natural way to do this is to use MC-SAT. However, it may be too expensive to run MC-SAT at every iteration of RMEU. It seems unlikely that flipping a single action grounding would affect a large number of clauses. The vast majority of clauses are unlikely to be affected by most flips; it is wasteful to repeatedly infer their probabilities at every iteration. Instead, we use an incremental version of MC-SAT to update only the probabilities of nodes significantly affected by the flip.

We initialize the probability distribution by running MC-SAT over the whole network before flipping any action groundings. After each flip, we run MC-SAT over a network containing the neighbors of the flipped node to see if they are significantly affected. We then add the neighbors of the affected nodes to the network, and run inference again. We repeat this until we find a network containing all the nodes significantly affected by the flip. Note that we do not need to calculate accurate probabilities during each iteration of incremental MC-SAT; we only need to know whether the probability has changed significantly since the last flip. Therefore, we only need a relatively small number of MC-SAT iterations for each iteration of incremental MC-SAT. Once we have found the set of nodes affected by the flip, we can run MC-SAT for more iterations to calculate accurate probabilities for those nodes.

FURTHER READING

Full details for these extensions can be found in their respective publications: hybrid Markov logic for handling continuous domains was introduced by Wang and Domingos [148]; Markov logic in infinite domains was introduced by Singla and Domingos [138]; recursive random fields were introduced by Lowd and Domingos [76]; and relational decision theory was introduced by Nath and Domingos [89].

CHAPTER 6

Applications

In this chapter, we go above the interface layer to show how Markov logic enables us to build state-of-the-art solutions for many problem domains. We cover eight applications: collective classification, social network analysis, entity resolution, information extraction, coreference resolution, robot mapping, link-based clustering, and semantic network extraction.

6.1 COLLECTIVE CLASSIFICATION

Collective classification is the task of inferring labels for a set of objects using not just their attributes but also the relations among them. For example, Web pages that link to each other tend to have similar topics. Markov logic makes it easy to incorporate relational information and jointly infer all of the labels at once.

A classic example of this is the WebKB dataset, which consists of labeled Web pages from the computer science departments of four universities. In this section, we show how to create a sophisticated model for collective classification on WebKB with just a few formulas in Markov logic. We used the relational version of the dataset from Craven and Slattery [16], which features 4165 Web pages and 10,935 Web links. Each Web page is marked with one of the following categories: student, faculty, professor, department, research project, course, or other. The goal is to predict these categories from the Web pages' words and links.

We can start with a simple logistic regression model, using only the words on the Web pages:

$$\text{PageClass}(p, +c)$$
$$\text{Has}(p, +w) \Rightarrow \text{PageClass}(p, +c)$$

The '+' operator here generates a separate rule (and with it, a separate learnable weight) for each constant of the appropriate type. The first line, therefore, generates a unit clause for each class, capturing the prior distribution over page classes. The second line generates thousands of rules representing the relationship between each word and each class. We can encode the fact that classes are mutually exclusive and exhaustive with a set of hard (infinite-weight) constraints:

$$\text{PageClass}(p, +c1) \wedge (+c1 \neq +c2) \Rightarrow \neg\text{PageClass}(p, +c2)$$
$$\exists c \, \text{PageClass}(p, c)$$

In Alchemy, we can instead state this property of the PageClass predicate in its definition using the '!' operator: PageClass(page, class!), where page and class are type names. (In general, the '!' notation signifies that, for each possible combination of values of the arguments without '!', there is exactly one combination of the arguments with '!' for which the predicate is true.)

To turn this multi-class logistic regression into a collective classification model with joint inference, we only need one more formula:

$$\texttt{Linked(u1, u2)} \land \texttt{PageClass(+c1, u1)} \land \texttt{PageClass(+c2, u2)}$$

This says that linked Web pages have related classes.

We performed leave-one-out cross-validation, training these models for 500 iterations of scaled conjugate gradient with a preconditioner (Section 4.1). The logistic regression baseline had an accuracy of 70.9%, while the model with joint inference had an accuracy of 76.4%. Markov logic makes it easy to construct additional features as well, such as words on linked pages, anchor text, etc. (See Taskar *et al.* [141] for a similar approach using relational Markov networks.)

6.2   SOCIAL NETWORK ANALYSIS AND LINK PREDICTION

In social network analysis, we seek to understand and reason about the connections among a group of people. A simple example of this is the friends and smokers domain introduced in Chapter 2. This example can easily be extended to include other attributes such as age, sex, etc. or other relationships such as coworker, parent, etc. Given such an MLN, we can predict relationships from the attributes (link prediction); attributes from the relationships (collective classification); or cluster entities based on their relationships (link-based clustering).

In addition to inferring specific facts, social network analysis is concerned with understanding the underlying causes and mechanisms that govern observed behavior. For such tasks, examining the learned weights can provide some intuition about which relationships and rules are most informative. For example, the weight learned for a transitivity clause (e.g., friends of friends are friends) says something about how cliquish a network is.

In the remainder of this section, we present a detailed example of link prediction in a social network, as done by Richardson and Domingos [117]. The specific task is to predict which professors advise which graduate students in the University of Washington Department of Computer Science and Engineering (UW-CSE). This dataset has been used as a benchmark for weight learning [133], structure learning [57, 83] and transfer learning [82]. We will focus, however, on the simplest scenario: generative weight learning with a manually specified knowledge base, using purely logical and probabilistic approaches as our baselines. This demonstrates both the applicability of Markov logic to social network analysis and its advantages over the pure approaches.

DATASET

The domain consists of 12 predicates and 2707 constants divided into 10 types. Types include: publication (342 constants), person (442), course (176), project (153), academic quarter (20), etc. Predicates include: `Professor(person)`, `Student(person)`, `AuthorOf(publication, person)`, `AdvisedBy(person, person)`, `YearsInProgram(person, years)`, `TeachingAssistant(course, person, quarter)`, `CourseLevel(course, level)`, `TaughtBy(course, person, quarter)`, etc.

Using typed variables, the total number of possible ground predicates is 4,106,841. The database contains a total of 3380 tuples (i.e., there are 3380 true ground predicates). We obtained this dataset by scraping pages in the department's Web site (http://www.cs.washington.edu).

To obtain a knowledge base, we asked four volunteers from the department to each provide a set of formulas in first-order logic describing the domain. Merging these yielded a KB of 96 formulas. Formulas in the KB include statements like: students are not professors; each student has at most one advisor; if a student is an author of a paper, so is her advisor; advanced students only TA courses taught by their advisors; at most, one author of a given publication is a professor; students in Phase I of the Ph.D. program have no advisor; etc. Notice that these statements are not always true, but are typically true.

For training and testing purposes, we divided the database into five sub-databases, one for each area: AI, graphics, programming languages, systems, and theory. We performed leave-one-out testing by area, testing on each area in turn using the model trained from the remaining four. The test task was to predict the AdvisedBy(x, y) predicate given (a) all others (All Info) and (b) all others except Student(x) and Professor(x) (Partial Info). In both cases, we measured the average conditional log-likelihood of all possible groundings of AdvisedBy(x, y) over all areas, drew precision/recall curves, and computed the area under the curve.

SYSTEMS
In order to evaluate MLNs, which use logic and probability for inference, we wished to compare with methods that use only logic or only probability. We were also interested in automatic induction of clauses using ILP techniques. This subsection gives details of the comparison systems used.

Logic
One important question we aimed to answer with the experiments is whether adding probability to a logical knowledge base improves its ability to model the domain. Doing this requires observing the results of answering queries using only logical inference, but this is complicated by the fact that computing log-likelihood and the area under the precision/recall curve requires real-valued probabilities, or at least some measure of "confidence" in the truth of each ground predicate being tested. We thus used the following approach. For a given knowledge base KB and set of evidence predicates E, let $X_{KB \cup E}$ be the set of worlds that satisfy $KB \cup E$. The probability of a query predicate q is then defined as $P(q) = \frac{X_{KB \cup E \cup q}}{X_{KB \cup E}}$, the fraction of $X_{KB \cup E}$ in which q is true.

A more serious problem arises if the KB is inconsistent (which was indeed the case with the KB collected from volunteers). In this case the denominator of $P(q)$ is zero. (Also, recall that an inconsistent knowledge base trivially entails any arbitrary formula). To address this, we redefined $X_{KB \cup E}$ to be the set of worlds that satisfies the maximum possible number of ground clauses. We used Gibbs sampling to sample from this set, with each chain initialized to a mode using WalkSat. At each Gibbs step, the step is taken with the following probability: 1 if the new state satisfies more clauses than the current one (since that means the current state should have 0 probability), 0.5 if the new state satisfies the same number of clauses (since the new and old state then have equal probability), and 0 if the new state satisfies fewer clauses. We then used only the states with maximum number of satisfied clauses to compute probabilities. Notice that this is equivalent to using an MLN built from the KB and with all infinite equal weights.

Probability

The other question we wanted to answer with these experiments is whether existing (propositional) probabilistic models are already powerful enough to be used in relational domains without the need for the additional representation power provided by MLNs. In order to use such models, the domain must first be propositionalized by defining features that capture useful information about it. Creating good attributes for propositional learners in this highly relational domain is a difficult problem. Nevertheless, as a tradeoff between incorporating as much potentially relevant information as possible and avoiding extremely long feature vectors, we defined two sets of propositional attributes: order-1 and order-2. The former involves characteristics of individual constants in the query predicate, and the latter involves characteristics of relations between the constants in the query predicate. For details on how these attributes were constructed, see Richardson and Domingos [117].

The resulting 28 order-1 attributes and 120 order-2 attributes (for the All Info case) were discretized into five equal-frequency bins (based on the training set). We used two propositional learners: Naive Bayes [30] and Bayesian networks [45] with structure and parameters learned using the VFBN2 algorithm [49] with a maximum of four parents per node. The order-2 attributes helped the naive Bayes classifier but hurt the performance of the Bayesian network classifier, so below we report results using the order-1 and order-2 attributes for naive Bayes, and only the order-1 attributes for Bayesian networks.

Inductive logic programming

The original knowledge base was acquired from volunteers, but we were also interested in whether it could have been developed automatically using inductive logic programming methods. We used CLAUDIEN [22] to induce a knowledge base from data. Besides inducing clauses from the training data, we were also interested in using data to automatically refine the knowledge base provided by our volunteers. CLAUDIEN does not support this feature directly, but it can be emulated by an appropriately constructed language bias.

MLNs

Our results compare the above systems to Markov logic networks. The MLNs were trained to maximize pseudo-log-likelihood, as described in Section 4.1 on generative weight learning. Since these experiments were performed before MC-SAT was developed, inference was done using Gibbs sampling. Typically, inference converged within 5000 to 100,000 passes.

RESULTS

We compared twelve systems: the original KB (KB); CLAUDIEN (CL); CLAUDIEN with the original KB as language bias (CLB); the union of the original KB and CLAUDIEN's output in both cases (KB+CL and KB+CLB); an MLN with each of the above KBs (MLN(KB), MLN(CL),

Table 6.1: Experimental results for predicting AdvisedBy(x, y) in the UW-CSE dataset when all other predicates are known (All Info) and when Student(x) and Professor(x) are unknown (Partial Info).

System	All Info		Partial Info	
	AUC	CLL	AUC	CLL
MLN(KB)	0.215: 0.0172	−0.052: 0.004	0.224: 0.0185	−0.048: 0.004
MLN(KB+CL)	0.152: 0.0165	−0.058: 0.005	0.203: 0.0196	−0.045: 0.004
MLN(KB+CLB)	0.011: 0.0003	−3.905: 0.048	0.011: 0.0003	−3.958: 0.048
MLN(CL)	0.035: 0.0008	−2.315: 0.030	0.032: 0.0009	−2.478: 0.030
MLN(CLB)	0.003: 0.0000	−0.052: 0.005	0.023: 0.0003	−0.338: 0.002
KB	0.059: 0.0081	−0.135: 0.005	0.048: 0.0058	−0.063: 0.004
KB+CL	0.037: 0.0012	−0.202: 0.008	0.028: 0.0012	−0.122: 0.006
KB+CLB	0.084: 0.0100	−0.056: 0.004	0.044: 0.0064	−0.051: 0.005
CL	0.048: 0.0009	−0.434: 0.012	0.037: 0.0001	−0.836: 0.017
CLB	0.003: 0.0000	−0.052: 0.005	0.010: 0.0001	−0.598: 0.003
NB	0.054: 0.0006	−1.214: 0.036	0.044: 0.0009	−1.140: 0.031
BN	0.015: 0.0006	−0.072: 0.003	0.015: 0.0007	−0.215: 0.003

MLN(KB+CL), and MLN(KB+CLB)); naive Bayes (NB); and a Bayesian network learner (BN). Add-one smoothing of probabilities was used in all cases.

Table 6.1 summarizes the results and Figure 6.1 shows precision/recall curves for all areas (i.e., averaged over all AdvisedBy(x, y) predicates). CLL is the average conditional log-likelihood, and AUC is the area under the precision-recall curve. (See http://www.cs.washington.edu/ai/mln for details on how the standard deviations of the AUCs were computed.) MLNs are clearly more accurate than the alternatives. The purely logical and purely probabilistic methods often suffer when intermediate predicates have to be inferred, while MLNs are largely unaffected. Naive Bayes performs well in AUC in some test sets, but very poorly in others; its CLLs are uniformly poor. CLAUDIEN performs poorly on its own, and produces no improvement when added to the KB in the MLN. Using CLAUDIEN to refine the KB typically performs worse in AUC but better in CLL than using CLAUDIEN from scratch; overall, the best-performing logical method is KB+CLB, but its results fall well short of the best MLNs'. The drop-off in precision around 50% recall is attributable to the fact that the database is very incomplete, and only allows identifying a minority of the AdvisedBy relations. Inspection reveals that the occasional smaller drop-offs in precision at very low recalls are due to students who graduated or changed advisors after co-authoring many publications with them.

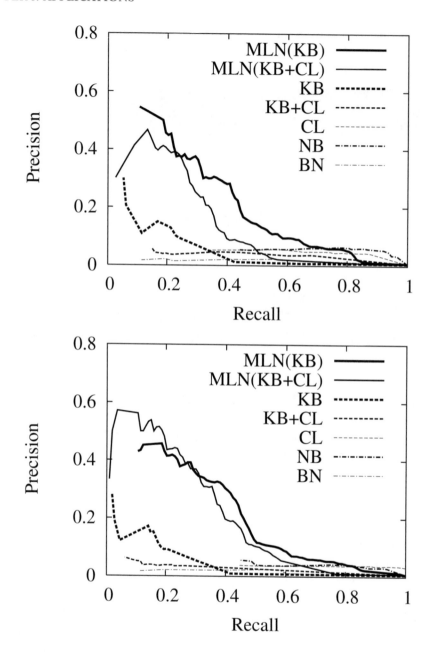

Figure 6.1: Precision and recall for link prediction in UW-CSE dataset: All Info (upper graph) and Partial Info (lower graph).

6.3   ENTITY RESOLUTION

Entity resolution is the problem of determining which observations (e.g., database records, noun phrases, video regions, etc.) correspond to the same real-world objects, and is of crucial importance in many areas. Typically, it is solved by forming a vector of properties for each pair of observations, using a learned classifier (such as logistic regression) to predict whether they match, and applying transitive closure. Markov logic yields an improved solution simply by applying the standard logical approach of removing the unique names assumption and introducing the equality predicate and its axioms: equality is reflexive, symmetric and transitive; groundings of a predicate with equal constants have the same truth values; and constants appearing in a ground predicate with equal constants are equal. This last axiom is not valid in logic, but captures a useful statistical tendency. For example, if two papers are the same, their authors are the same; and if two authors are the same, papers by them are more likely to be the same. Weights for different instances of these axioms can be learned from data. Inference over the resulting MLN, with entity properties and relations as the evidence and equality atoms as the query, naturally combines logistic regression and transitive closure. Most importantly, it performs *collective* entity resolution, where resolving one pair of entities helps to resolve pairs of related entities.

As a concrete example, consider the task of deduplicating a citation database in which each citation has author, title, and venue fields, as done by Singla and Domingos [134]. We can represent the domain structure with eight relations: Author(bib, author), Title(bib, title), and Venue(bib, venue) relate citations to their fields; HasWord(author/title/ venue, word) indicates which words are present in each field; SameAuthor (author, author), SameTitle(title, title), and SameVenue(venue, venue) represent field equality; and SameBib(bib, bib) represents citation equality. The truth values of all relations except for the equality relations are provided as evidence. The objective is to predict the SameBib relation.

We begin with a logistic regression model to predict citation equality based on the words in the fields. This is easily expressed in Markov logic by rules such as the following:

$$\text{Title}(b1, t1) \quad \wedge \text{Title}(b2, t2) \wedge \text{HasWord}(t1, +\text{word})$$
$$\wedge \text{HasWord}(t2, +\text{word}) \Rightarrow \text{SameBib}(b1, b2)$$

When given a positive weight, each of these rules increases the probability that two citations with a particular title word in common are equivalent. We can construct similar rules for other fields. Note that we may learn negative weights for some of these rules, just as logistic regression may learn negative feature weights. Transitive closure consists of a single rule:

$$\text{SameBib}(b1, b2) \wedge \text{SameBib}(b2, b3) \Rightarrow \text{SameBib}(b1, b3)$$

This model is similar to the standard solution, but has the advantage that the classifier is learned in the context of the transitive closure operation.

We can construct similar rules to predict the equivalence of two fields as well. The usefulness of Markov logic is shown further when we link field equivalence to citation equivalence:

$$\text{Author}(b1, a1) \land \text{Author}(b2, a2) \land \text{SameBib}(b1, b2) \Rightarrow \text{SameAuthor}(a1, a2)$$
$$\text{Author}(b1, a1) \land \text{Author}(b2, a2) \land \text{SameAuthor}(a1, a2) \Rightarrow \text{SameBib}(b1, b2)$$

The above rules state that if two citations are the same, their authors should be the same, and that citations with the same author are more likely to be the same.

Most importantly, the resulting model can now perform *collective* entity resolution, where resolving one pair of entities helps to resolve pairs of related entities. For example, inferring that a pair of citations are equivalent can provide evidence that the names *AAAI-06* and *21st Natl. Conf. on AI* refer to the same venue, even though they are superficially very different. This equivalence can then aid in resolving other entities.

Experiments on citation databases like Cora and BibServ.org show that these methods can greatly improve accuracy, particularly for entity types that are difficult to resolve in isolation, as in the above example [134]. Due to the large number of words and the high arity of the transitive closure formula, these models have thousands of weights and ground millions of clauses during learning, even after using canopies to limit the number of comparisons considered. Learning at this scale is still reasonably efficient: preconditioned scaled conjugate gradient with MC-SAT for inference converges within a few hours [75]. Rather than presenting detailed experimental results for entity resolution, we will show in the next section how to incorporate automatic field segmentation and then compare the performance of the full system to several state-of-the-art methods.

6.4   INFORMATION EXTRACTION

In the citation example of the previous section, it was assumed that the fields were manually segmented in advance. The goal of information extraction is to extract database records starting from raw text or semi-structured data sources. Traditionally, information extraction proceeds by first segmenting each candidate record separately, and then merging records that refer to the same entities. Such a pipeline architecture is adopted by many AI systems in natural language processing, speech recognition, vision, robotics, etc. Markov logic allows us to perform the two tasks jointly. This enables us to use the segmentation of one candidate record to help segment similar ones. For example, resolving a well-segmented field with a less-clear one can disambiguate the latter's boundaries. We will continue with the example of citations, as done by Poon and Domingos [106], but similar ideas could be applied to other data sources, such as Web pages or emails.

The main evidence predicate in the information extraction MLN is $\text{Token}(t, i, c)$, which is true iff token t appears in the ith position of the cth citation. A token can be a word, date, number, etc. Punctuation marks are not treated as separate tokens; rather, the predicate $\text{HasPunc}(c, i)$ is true iff a punctuation mark appears immediately after the ith position in the cth citation. The query predicates are $\text{InField}(i, f, c)$ and $\text{SameCitation}(c, c')$. $\text{InField}(i, f, c)$ is true iff the ith position of the cth citation is part of field f, where $f \in \{\text{Title}, \text{Author}, \text{Venue}\}$, and inferring it performs segmentation. SameCitation(c, c') is true iff citations c and c' represent the same publication, and inferring it performs entity resolution.

Our segmentation model is essentially a hidden Markov model (HMM) with enhanced ability to detect field boundaries. The observation matrix of the HMM correlates tokens with fields, and is represented by the simple rule

$$\text{Token}(+t, i, c) \Rightarrow \text{InField}(i, +f, c)$$

If this rule was learned in isolation, the weight of the (t, f)th instance would be $\log(p_{tf}/(1 - p_{tf}))$, where p_{tf} is the corresponding entry in the HMM observation matrix. In general, the transition matrix of the HMM is represented by a rule of the form

$$\text{InField}(i, +f, c) \Rightarrow \text{InField}(i + 1, +f', c)$$

However, we (and others, e.g., [41]) have found that for segmentation it suffices to capture the basic regularity that consecutive positions tend to be part of the same field. Thus we replace f' by f in the formula above. We also impose the condition that a position in a citation string can be part of at most one field; it may be part of none.

The main shortcoming of this model is that it has difficulty pinpointing field boundaries. Detecting these is key for information extraction, and a number of approaches use rules designed specifically for this purpose (e.g., [63]). In citation matching, boundaries are usually marked by punctuation symbols. This can be incorporated into the MLN by modifying the rule above to

$$\text{InField}(i, +f, c) \wedge \neg \text{HasPunc}(c, i) \Rightarrow \text{InField}(i + 1, +f, c)$$

The $\neg\text{HasPunc}(c, i)$ precondition prevents propagation of fields across punctuation marks. Because propagation can occur differentially to the left and right, the MLN also contains the reverse form of the rule. In addition, to account for commas being weaker separators than other punctuation, the MLN includes versions of these rules with HasComma() instead of HasPunc().

Finally, the MLN contains rules capturing a variety of knowledge about citations: the first two positions of a citation are usually in the author field, and the middle one in the title; initials (e.g., "J.") tend to appear in either the author or the venue field; positions preceding the last non-venue initial are usually not part of the title or venue; and positions after the first venue keyword (e.g., "Proceedings", "Journal") are usually not part of the author or title.

By combining this segmentation model with our entity resolution model from before, we can exploit relational information as part of the segmentation process. In practice, something a little more sophisticated is necessary to get good results on real data. In Poon and Domingos [106], we define predicates and rules specifically for passing information between the stages, as opposed to just using the existing InField() outputs. This leads to a "higher bandwidth" of communication between segmentation and entity resolution, without letting excessive segmentation noise through. We also define an additional predicate and modify rules to better exploit information from similar citations during the segmentation process. See Poon and Domingos [106] for further details.

Table 6.2: CiteSeer entity resolution: cluster recall on each section.

Approach	Constr.	Face	Reason.	Reinfor.
Fellegi-Sunter	84.3	81.4	71.3	50.6
Lawrence *et al.* (1999)	89	94	86	79
Pasula *et al.* (2002)	93	97	96	94
Wellner *et al.* (2004)	95.1	96.9	93.7	94.7
Joint MLN	96.0	97.1	95.1	96.7

Table 6.3: Cora entity resolution: pairwise recall/precision and cluster recall.

Approach	Pairwise Rec./Prec.	Cluster Recall
Fellegi-Sunter	78.0 / 97.7	62.7
Joint MLN	94.3 / 97.0	78.1

We evaluated this model on the CiteSeer and Cora datasets. For entity resolution in Cite-Seer, we measured *cluster recall* for comparison with previously published results. Cluster recall is the fraction of clusters that are correctly output by the system after taking transitive closure from pairwise decisions. For entity resolution in Cora, we measured both cluster recall and pairwise recall/precision. In both datasets, we also compared with a "standard" Fellegi-Sunter model (see Singla and Domingos [134]), learned using logistic regression, and with oracle segmentation as the input.

In both datasets, joint inference improved accuracy and our approach outperformed previous ones. Table 6.2 shows that our approach outperforms previous ones on CiteSeer entity resolution. (Results for Lawrence *et al.* (1999) [66], Pasula *et al.* (2002) [98] and Wellner *et al.* (2004) [152] are taken from the corresponding papers.) This is particularly notable given that the models of Pasula *et al.* [98] and Wellner *et al.* [152] involved considerably more knowledge engineering than ours, contained more learnable parameters, and used additional training data.

Table 6.3 shows that our entity resolution approach easily outperforms Fellegi-Sunter on Cora, and has very high pairwise recall/precision.

6.5 UNSUPERVISED COREFERENCE RESOLUTION

In natural language processing, the goal of coreference resolution is to identify *mentions* (typically noun phrases) within a document that refer to the same *entities*. This is a key subtask in many NLP applications, including information extraction, question answering, machine translation, and others. Supervised learning approaches treat the problem as one of classification: for each pair of mentions, predict whether they corefer or not (e.g., McCallum and Wellner [80]). While successful, these approaches require labeled training data, consisting of mention pairs and the correct decisions for them. This limits their applicability. Unsupervised approaches are attractive due to the availability of large quantities of unlabeled text. However, unsupervised coreference resolution is much more difficult. Haghighi and Klein's model [42], the most sophisticated prior work, still lags behind supervised ones by a substantial margin.

The lack of label information in unsupervised coreference resolution can potentially be overcome by performing joint inference, which leverages the "easy" decisions to help make related "hard" ones. Markov logic allows us to easily build models involving relations among mentions, like apposition and predicate nominals. In this section, we describe Poon and Domingos' method for joint unsupervised coreference resolution using Markov logic [107]. On the standard MUC and ACE datasets, this approach outperforms Haghighi and Klein's using only a fraction of the training data, and often matches or exceeds the accuracy of state-of-the-art supervised models.

AN MLN FOR JOINT UNSUPERVISED COREFERENCE RESOLUTION

In this subsection, we present an MLN for joint unsupervised coreference resolution. As common in previous work, we assume that true mention boundaries are given. We do not assume any other labeled information. We determined the head of a mention either by taking its rightmost token, or by using the head rules in a parser. We detected pronouns using a list.

Base MLN

The main query predicate is InClust(m, c!), which is true iff mention m is in cluster c. The "c!" notation signifies that for each m, this predicate is true for a unique value of c. The main evidence predicate is Head(m, t!), where m is a mention and t a token, and which is true iff t is the head of m. A key component in our MLN is a simple head mixture model, where the mixture component priors are represented by the unit clause InClust(+m, +c) and the head distribution is represented by the *head prediction rule*

$$InClust(m, +c) \land Head(m, +t).$$

By convention, at each inference step we name each non-empty cluster after the earliest mention it contains. This helps break the symmetry among mentions, which otherwise produces multiple optima and makes learning unnecessarily harder. To encourage clustering, we impose an exponential prior on the number of non-empty clusters with weight -1.

The above model only clusters mentions with the same head, and does not work well for pronouns. To address this, we introduce the predicate IsPronoun(m), which is true iff the mention m is a pronoun, and adapt the head prediction rule as follows:

$$\neg IsPronoun(m) \land InClust(m, +c) \land Head(m, +t)$$

This is always false when m is a pronoun, and thus applies only to non-pronouns.

Pronouns tend to resolve with mentions that are semantically compatible with them. Thus we introduce predicates that represent entity type, number, and gender: Type(x, e!), Number(x, n!), Gender(x, g!), where x can be either a cluster or mention, e ∈ {Person, Organization, Location, Other}, n ∈ {Singular, Plural} and g ∈

{Male, Female, Neuter}. Many of these are known for pronouns, and some can be inferred from simple linguistic cues (e.g., "Ms. Galen" is a singular female person, while "XYZ Corp." is an organization). Entity type assignment is represented by the unit clause Type(+x, +e) and similarly for number and gender. A mention should agree with its cluster in entity type. This is ensured by the hard (infinite-weight) rule

$$\text{InClust}(m, c) \Rightarrow (\text{Type}(m, e) \Leftrightarrow \text{Type}(c, e))$$

There are similar hard rules for number and gender.

Different pronouns prefer different entity types, as represented by

$$\text{IsPronoun}(m) \wedge \text{InClust}(m, c) \wedge \text{Head}(m, +t) \wedge \text{Type}(c, +e)$$

which only applies to pronouns, and whose weight is positive if pronoun t is likely to assume entity type e and negative otherwise. There are similar rules for number and gender.

Aside from semantic compatibility, pronouns tend to resolve with nearby mentions. To model this, we impose an exponential prior on the distance (number of mentions) between a pronoun and its antecedent, with weight -1.

Full MLN

Syntactic relations among mentions often suggest coreference. Incorporating such relations into our MLN is straightforward. We illustrate this with two examples: apposition (e.g., "Morgan & Claypool, our publisher") and predicate nominals (e.g., "Morgan & Claypool is our publisher"). We introduce a predicate for apposition, Appo(x, y), where x, y are mentions, and which is true iff y is an appositive of x. We then add the rule

$$\text{Appo}(x, y) \Rightarrow (\text{InClust}(x, c) \Leftrightarrow \text{InClust}(y, c))$$

which ensures that x, y are in the same cluster if y is an appositive of x. Similarly, we introduce a predicate for predicate nominals, PredNom(x, y), and the corresponding rule.[1] The weights of both rules can be learned from data with a positive prior mean. For simplicity, we treat them as hard constraints.

Rule-Based MLN

We also consider a rule-based system that clusters non-pronouns by their heads, and attaches a pronoun to the cluster which has no known conflicting type, number, or gender, and contains the closest antecedent for the pronoun. This system can be encoded in an MLN with just four rules. Three of them are the ones for enforcing agreement in type, number, and gender between a cluster and its members, as defined in the base MLN. The fourth rule is

$$\neg\text{IsPronoun}(m1) \wedge \neg\text{IsPronoun}(m2) \wedge \text{Head}(m1, h1) \wedge \text{Head}(m2, h2)$$
$$\wedge\text{InClust}(m1, c1) \wedge \text{InClust}(m2, c2) \Rightarrow (c1 = c2 \Leftrightarrow h1 = h2).$$

---

[1]We detected apposition and predicate nominatives using simple heuristics based on parses, e.g., if (NP, comma, NP) are the first three children of an NP, then any two of the three noun phrases are apposition.

With a large but not infinite weight (e.g., 100), this rule has the effect of clustering non-pronouns by their heads, except when it violates the hard rules. The MLN can also include the apposition and predicate-nominal rules. As in the base MLN, we impose the same exponential prior on the number of non-empty clusters and that on the distance between a pronoun and its antecedent. This simple MLN is remarkably competitive, as we will see below.

EXPERIMENTS

We conducted experiments on MUC-6, ACE-2004, and ACE Phrase-2 (ACE-2). We evaluated our systems using the commonly-used MUC scoring program [147].

The MUC-6 dataset consists of 30 documents for testing and 221 for training. To evaluate the contribution of the major components in our model, we conducted five experiments, each differing from the previous one in a single aspect. We emphasize that our approach is unsupervised, and thus the data only contains raw text plus true mention boundaries.

MLN-1 In this experiment, the base MLN was used, and the head was chosen crudely as the rightmost token in a mention. Our system was run on each test document separately, using a minimum of training data (the document itself).

MLN-30 Our system was trained on all 30 test documents together. This tests how much can be gained by pooling information.

MLN-H The heads were determined using the head rules in the Stanford parser [56], plus simple heuristics to handle suffixes such as "Corp." and "Inc."

MLN-HA The apposition rule was added.

MLN-HAN The predicate-nominal rule was added. This is our full model.

We also compared with two rule-based MLNs: RULE chose the head crudely as the rightmost token in a mention, and did not include the apposition rule and predicate-nominal rule; RULE-HAN chose the head using the head rules in the Stanford parser, and included the apposition rule and predicate-nominal rule.

We used the ACE Phrase-2 dataset (ACE-2) to enable direct comparisons with Haghighi and Klein [42], Ng [92], and Denis and Baldridge [27]. The English version of the ACE-2004 training corpus contains two sections, BNEWS and NWIRE, with 220 and 128 documents, respectively. ACE-2 contains a training set and a test set. In our experiments, we only used the test set, which contains three sections, BNEWS, NWIRE, and NPAPER, with 51, 29, and 17 documents, respectively.

We applied the weight learning methods from Section 4.3 to maximize the likelihood of the observed query atoms given the evidence atoms, while summing over the unobserved query atoms: $P(y_o|x) = \sum_{y_u} P(y_o, y_u|x)$. In our coreference resolution MLN, known query atoms (Y_o) include Head and known groundings of Type, Number and Gender; unknown query atoms (Y_u)

Table 6.4: Comparison of coreference results in MUC scores on the MUC-6 dataset.

	# Doc.	Prec.	Rec.	F1
H&K	60	80.8	52.8	63.9
H&K	381	80.4	62.4	70.3
M&W	221	-	-	73.4
RULE	-	76.0	65.9	70.5
RULE-HAN	-	81.3	72.7	76.7
MLN-1	1	76.5	66.4	71.1
MLN-30	30	77.5	67.3	72.0
MLN-H	30	81.8	70.1	75.5
MLN-HA	30	82.7	75.1	78.7
MLN-HAN	30	83.0	75.8	79.2

Table 6.5: Comparison of coreference results in MUC scores on the ACE-2 datasets.

	BNEWS			NWIRE			NPAPER		
	Prec.	Rec.	F1	Prec.	Rec.	F1	Prec.	Rec.	F1
Ng	67.9	62.2	64.9	60.3	50.1	54.7	71.4	67.4	69.3
D&B	78.0	62.1	69.2	75.8	60.8	67.5	77.6	68.0	72.5
MLN-HAN	68.3	66.6	67.4	67.7	67.3	67.4	69.2	71.7	70.4

include `InClust` and unknown groundings of `Type`, `Number`, `Gender`; and evidence (X) includes `IsPronoun`, `Appo` and `PredNom`. For inference, we adapted MC-SAT and MaxWalkSAT to better handle mutually exclusive and exhaustive rules. Additional details can be found in Poon and Domingos [107].

Table 6.4 compares our system with previous approaches on the MUC-6 dataset, in MUC scores. Our approach greatly outperformed Haghighi and Klein, the previous state-of-the-art unsupervised system. Our system, when trained on individual documents, achieved an F1 score more than 7% higher than theirs trained on 60 documents, and still outperformed it when trained on 381 documents. Training on the 30 test documents together resulted in a significant gain. (We also ran experiments using more documents, and the results were similar.) Better head identification (MLN-H) led to a large improvement in accuracy, which is expected since for mentions with a right modifier, the rightmost tokens confuse rather than help coreference (e.g., "the chairman of Microsoft"). Notice that with this improvement our system already outperforms a state-of-the-art supervised system, McCallum and Wellner [80]. Leveraging apposition resulted in another large improvement, and predicate nominals also helped. Our full model scores about 9% higher than Haghighi and Klein, and about 6% higher than McCallum and Wellner. To our knowledge, this is the best coreference accuracy reported on MUC-6 to date.[2] Interestingly, the rule-based MLN (RULE) sufficed to outperform Haghighi and Klein, and by using better heads and the apposition and predicate-nominal rules (RULE-HAN), it outperformed McCallum and Wellner, the supervised system. The MLNs with learning (MLN-30 and MLN-HAN), on the other hand, substantially outperformed the corresponding rule-based ones.

Tables 6.5 compares our system to two recent supervised systems, Ng [92] and Denis and Baldridge [27]. Our approach significantly outperformed Ng. It tied with Denis and Baldridge on NWIRE, and was somewhat less accurate on BNEWS and NPAPER.

6.6   ROBOT MAPPING

In robot mapping, the goal is to infer the map of an indoor environment from laser range data, obtained by the robot as it moves about the environment [69]. Since this domain relies heavily on real numbers and continuous variables for distances and coordinates, it is best approached using hybrid Markov logic networks (HMLNs), as described in Section 5.1.

Using the mobile robot sensor data from the Radish robotics data set repository (http://radish.sourceforge.net), the evidence is a set of range finder segments, defined by the (x, y) coordinates of their endpoints. The output is: (a) a labeling of each segment as Door, Wall, or Other (classification); (b) an assignment of wall segments to the walls they are part of (clustering); and (c) the position of each wall, defined by the (x, y) coordinates of its endpoints (regression). On average, each map consists of about 100 segments. More details on the dataset and experimental setups can be found in Wang and Domingos [148].

Wang and Domingos [148] built an HMLN using the following rules. First, every segment belongs to exactly one type with some prior probability distribution given by the unit clauses, SegType(s, +t). Doors and walls also have a typical length and depth, represented by numeric terms (e.g., SegType(s, Door) · (Length(s) = DoorLength), which has value 0 if the segment is not a door and $-(\text{Length(s)} - \text{DoorLength})^2$ otherwise). A segment's depth is defined as the signed perpendicular distance of its midpoint to the nearest wall line. Lines are iteratively estimated from the segments assigned to them, using the formulas below, as part of the HMLN inference. A number of rules involving depth and angle between a segment and the nearest line identify segments of type Other, whose distribution is more irregular than that of doors and walls.

The type of a segment is predictive of the types of consecutive segments along a wall line:

$$\text{SegType(s, +t)} \wedge \text{Consecutive(s, s}') \Rightarrow \text{SegType(s}', +t')$$

Two segments are consecutive if they are the closest segments to each other along either direction of the nearest line. In addition, aligned segments tend to be of the same type:

$$\text{SegType(s, t)} \wedge \text{Consecutive(s, s}') \wedge \text{Aligned(s, s}') \Rightarrow \text{SegType(s}', t)$$

---

[2] As pointed out by Haghighi and Klein [42], Luo *et al.* [77] obtained a very high accuracy on MUC-6, but their system used gold NER features and is not directly comparable.

Segments are aligned if one is roughly a continuation of the other (i.e., their angle is below some threshold, and so is their perpendicular distance). The rules above perform collective classification of segments into types.

Aligned wall segments are part of the same wall line:

$$\text{SegType}(s, \text{Wall}) \wedge \text{SegType}(s', \text{Wall}) \wedge \text{Aligned}(s, s') \wedge \text{PartOf}(s, l) \Rightarrow \text{PartOf}(s', l)$$

This rule clusters wall segments into wall lines. Notice that it captures long-range dependencies between segments along the same corridor, not just between neighboring segments.

A segment is the start of a line if and only if it has no previous aligned segment:[3]

$$\text{PartOf}(s, l) \Rightarrow (\neg \text{PreviousAligned}(s) \Leftrightarrow \text{StartLine}(s, l))$$

If a segment is the start of a line, their initial points are (about) the same:

$$\text{StartLine}(s, l) \cdot (x_i(s) = x_i(l))$$

where $x_i(s)$ is the x coordinate of s's initial point, etc.; and similarly for y. If a segment belongs to a line, its slope should be the same as the slope of a line connecting their initial points. Multiplying by the Δx's to avoid singularities, we obtain:

$$\text{PartOf}(s, l) \cdot [(y_f(s) - y_i(s))(x_i(s) - x_i(l)) = (y_i(s) - y_i(l))(x_f(s) - x_i(s))]$$

where the subscript f denotes final points. Line ends are handled similarly. These rules infer the locations of the wall lines from the segments assigned to them. They also influence the clustering and labeling of segments (e.g., if a better line fit is obtained by relabeling a segment from Wall to Door and thus excluding it, this will be inferred). Classification, clustering and regression are fully joint. In the next section we show the experimental results obtained by this approach. The full HMLN, containing some additional formulas, is available at the Alchemy Web site (http://alchemy.cs.washington.edu/).

EXPERIMENTAL METHODOLOGY

We learned HMLN weights using Alchemy's voted perceptron with HMWS for inference, 100 steps of gradient descent, and a learning rate of 1.0. To combat ill-conditioning, we divided each formula/term's learning rate by the absolute sum of its values in the data, similar to per-weight learning rates discussed in Section 4.1. Other parameters were learned by fitting Gaussian distributions to the relevant variables. We used leave-one-map-out cross-validation throughout. In all inferences, the correct number of walls was given. We evaluated inference results for discrete variables (SegType and PartOf) using F1 score (harmonic mean of precision and recall over all groundings). We used mean square error (MSE) for continuous variables ($x_i(l)$, $y_i(l)$, $x_f(l)$ and $y_f(l)$). We also computed the

---

[3]By convention, a point precedes another if it has lower x coordinate and, if they are the same, lower y. Segments and lines are ordered by their start points, which precede their final points.

Table 6.6: Joint vs. pipeline approaches for robot mapping domain.

Measure	SegType F1	PartOf F1	MSE
HMWS	0.746	0.753	0.112
MWS+LR	0.742	0.717	0.198
HMCS	0.922	0.931	0.002
MCS+LR	0.904	0.919	0.037
RMNs	0.899	N/A	N/A

negative log likelihoods of the test values. To obtain density functions for continuous variables from the output of MCMC, we placed a Gaussian kernel at each sample, with a standard deviation of $3r/n$, where r is the sampled range of the variable and n is the number of samples. Details of the experimental procedure, parameter settings and results can be found in Wang and Domingos [148] and the associated online appendix.

INFERRING THE MOST PROBABLE STATE

We compared HMWS, which infers the discrete and continuous variables jointly, with a more traditional pipeline approach, where the segment types and assignments are first inferred using MWS, and the lines are then estimated by least-squares linear regression over the endpoints of the segments assigned to them by MWS. HMWS and MWS were started from the same initial random state and run for 1,000,000 flips; they typically converged within 100,000 flips. The results are shown in the top two rows of Table 6.6. Joint inference outperforms pipelined inference on both discrete and numeric variables.

INFERRING CONDITIONAL PROBABILITIES

We compared HMCS with a pipeline approach, where the discrete variables are first inferred using MC-SAT, and the lines are then estimated by linear regression over the endpoints of the segments that have probability greater than 0.3 of belonging to them. (There are typically four to six lines in a map.) HMCS and MC-SAT were both run for 30,000 steps, with all other settings as above. The results are shown in Table 6.6 (the bottom three lines). Joint inference performs best, illustrating its benefits. Table 6.6 also shows the results on classifying SegType of Limketkai *et al.*'s state-of-the-art approach [69], which is based on relational Markov networks [141]. HMLNs perform best, illustrating the benefits of using a more flexible modeling language.

6.7   LINK-BASED CLUSTERING

Clustering is the task of automatically partitioning a set of objects into groups so that objects within each group are similar to each other while being dissimilar to those in other groups. Link-based clustering uses relationships among the objects in determining similarity. We can do this very naturally with the MRC algorithm from Section 4.3, which is based on second-order Markov logic. Note that second-order formulas are quantified over predicates, so the same second-order MLN can be applied to different datasets with different predicates. In this section, we describe Kok and Domingos' [58] experiments with MRC on four datasets. We compare MRC to the infinite relational model (IRM) [54] and TDSL, the top-down structure learning algorithm presented in Section 4.2.

IRM is a recently-published model that also clusters objects, attributes, and relations. However, unlike MRC, it only finds a single clustering. It defines a generative model for the predicates and cluster assignments. Like MRC, it assumes that the predicates are conditionally independent given the cluster assignments, and the cluster assignments for each type are independent. IRM uses a Chinese restaurant process prior (CRP) [102] on the cluster assignments. Under the CRP, each new object is assigned to an existing cluster with probability proportional to the cluster size. Because the CRP has the property of exchangeability, the order in which objects arrive does not affect the outcome. IRM assumes that the probability p of an atom being true conditioned on cluster membership is generated according to a symmetric Beta distribution, and that the truth values of atoms are then generated according to a Bernoulli distribution with parameter p. IRM finds the MAP cluster assignment using the same greedy search as our model, except that it also searches for the optimal values of its CRP and Beta parameters.

DATASETS

Animals. This dataset contains a set of animals and their features [96]. It consists exclusively of unary predicates of the form f(a), where f is a feature and a is an animal (e.g., Swims(Dolphin)). There are 50 animals, 85 features, and thus a total of 4250 ground atoms, of which 1562 are true. This is a simple propositional dataset with no relational structure, but it is useful as a "base case" for comparison. Notice that, unlike traditional clustering algorithms, which only cluster objects by features, MRC and IRM also cluster features by objects. This is known as bi-clustering or co-clustering, and has received considerable attention in the recent literature (e.g., Dhillon *et al.* [28]).

UMLS. UMLS contains data from the Unified Medical Language System, a biomedical ontology [81]. It consists of binary predicates of the form r(c, c'), where c and c' are biomedical concepts (e.g., Antibiotic, Disease), and r is a relation between them (e.g., Treats, Diagnoses). There are 49 relations and 135 concepts, for a total of 893,025 ground atoms, of which 6529 are true.

Kinship. This dataset contains kinship relationships among members of the Alyawarra tribe from Central Australia [26]. Predicates are of the form k(p, p'), where k is a kinship relation and p, p' are persons. There are 26 kinship terms and 104 persons, for a total of 281,216 ground atoms, of which 10,686 are true.

Nations. This dataset contains a set of relations among nations and their features [126]. It consists of binary and unary predicates. The binary predicates are of the form r(n, n'), where n, n' are nations, and r is a relation between them (e.g., ExportsTo, GivesEconomicAidTo). The unary predicates are of the form `f(n)`, where `n` is a nation and `f` is a feature (e.g., `Communist`, `Monarchy`). There are 14 nations, 56 relations and 111 features, for a total of 12,530 ground atoms, of which 2565 are true.

METHODOLOGY

Experimental evaluation of statistical relational learners is complicated by the fact that in many cases the data cannot be separated into independent training and test sets. While developing a long-term solution for this remains an open problem, we used an approach that is general and robust: performing cross-validation by atom. For each dataset, we performed ten-fold cross-validation by randomly dividing the atoms into ten folds, training on nine folds at a time, and testing on the remaining one. This can be seen as evaluating the learners in a *transductive* setting, because an object (e.g., `Leopard`) that appears in the test set (e.g., in `MeatEater(Leopard)`) may also appear in the training set (e.g., in `Quadrupedal(Leopard)`). In the training data, the truth values of the test atoms are set to `unknown`, and their actual values (`true/false`) are not available. Thus learners must perform generalization in order to be able to infer the test atoms, but the generalization is aided by the dependencies between test atoms and training ones.

TDSL is not directly comparable to MRC and IRM because it makes the closed-world assumption and our experiments require the test atoms to be open-world. For an approximate comparison, we set all test atoms to false when running TDSL. Since in each run these are only 10% of the training set, setting them to false does not greatly change the sufficient statistics (true clause counts). We then ran MC-SAT on the MLNs learned by TDSL to infer the probabilities of the test atoms.

We ran IRM and MRC for ten hours on each fold of each dataset, using default parameter settings for both algorithms. Almost all of MRC's running time went into the first level of clustering because this is where the sets of objects, attributes and relations to be clustered are by far the largest, and finding a good initial clustering is important for the subsequent learning.

RESULTS

Figure 6.2 reports the CLL and AUC for MRC, IRM and TDSL, averaged over the ten folds of each dataset. We also report the results obtained using just the initial clustering formed by MRC (Init), in order to evaluate the usefulness of learning multiple clusterings.

TDSL does worse than MRC and IRM on all datasets except Nations. On Nations, it does worse than MRC and IRM in terms of CLL, but approximately ties them in terms of AUC. Many of the relations in Nations are symmetric, e.g., if country A has a military conflict with B, then the reverse is usually true. TDSL learns a rule to capture the symmetry, and consequently does well in terms of AUC.

MRC outperforms IRM on UMLS and Kinship, and ties it on Animals and Nations. The difference on UMLS and Kinship is quite large. Animals is the smallest and least structured of the datasets, and it is conceivable that it has little room for improvement beyond a single clustering. The difference in performance between MRC and IRM correlates strongly with dataset size. (Notice

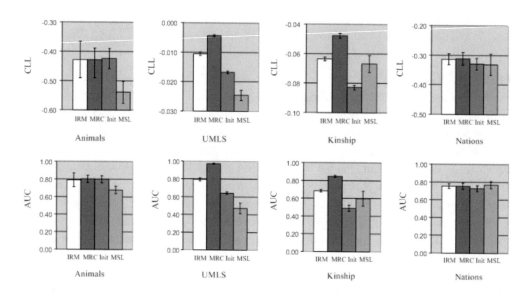

Figure 6.2: Comparison of MRC, IRM, and MLN structure learning (TDSL) on four clustering datasets. Init is the initial clustering formed by MRC. Error bars are one standard deviation in each direction.

that UMLS and Kinship are an order of magnitude larger than Animals and Nations.) This suggests that sophisticated algorithms for statistical predicate invention may be of most use in even larger datasets. We look at such a dataset in the next section.

MRC outperforms Init on all domains except Animals. The differences on Nations are not significant, but on UMLS and Kinship they are very large. These results show that forming multiple clusterings is key to the good performance of MRC. In fact, Init does considerably worse than IRM on UMLS and Kinship; we attribute this to the fact that IRM performs a search for optimal parameter values, while in MRC these parameters were simply set to default values without any tuning on data. This suggests that optimizing parameters in MRC could lead to further performance gains.

In the Animals dataset, MRC performs at most three levels of cluster refinement. On the other datasets, it performs about five. The average total numbers of clusters generated are: Animals, 202; UMLS, 405; Kinship, 1044; Nations, 586. The average numbers of atom prediction rules learned are: Animals, 305; UMLS, 1935; Kinship, 3568; Nations, 12,169.

6.8   SEMANTIC NETWORK EXTRACTION FROM TEXT

A long-standing goal of AI is to build an autonomous agent that can read and understand text. A good step towards this goal is automatically building knowledge bases of facts and relationships from corpora such as the Web. Kok and Domingos [59] recently introduced Semantic Network Extractor

(SNE), a scalable, unsupervised, and domain-independent system that simultaneously extracts high-level relations and concepts, and learns a semantic network [112] from text. The knowledge in this network can be used by itself or as a component of larger NLP systems. SNE represents the largest-scale application of Markov logic to date.

SNE is built on top of TextRunner [3], a system for extracting a large set of relational tuples of the form r(x, y) where x and y are strings denoting objects, and r is a string denoting a relation between the objects. TextRunner uses a lightweight noun phrase chunker to identify objects, and heuristically determines the text between objects as relations. However, while TextRunner can quickly acquire a large database of ground facts in an unsupervised manner, it is not able to learn general knowledge that is embedded in the facts.

SNE uses TextRunner to extract ground facts as triples from text, and then extract knowledge from the triples. TextRunner's triples are noisy, sparse, and contain many co-referent objects and relations. SNE overcomes this by using a probabilistic model based on second-order Markov logic that clusters objects according to the objects they are related to and relations according to the objects they relate. This allows information to propagate between clusters of relations and clusters of objects as they are created. Each cluster represents a high-level relation or concept. A concept cluster can be viewed as a node in a graph, and a relation cluster can be viewed as links between the concept clusters that it relates. Together, the concept clusters and relation clusters define a simple semantic network. Figure 6.3 illustrates part of a semantic network that our approach learns. (More fragments are available at http://alchemy.cs.washington.edu/papers/kok08.) SNE is based on the MRC model [58] discussed in Section 4.3, but with a different prior and a more scalable search algorithm that finds a single clustering. For full details on SNE, see Kok and Domingos [59].

Our goal is to create a system that is capable of extracting semantic networks from what is arguably the largest and most accessible text resource — the Web. Thus in our experiments, we used a large Web corpus to evaluate the effectiveness of SNE's relational clustering approach in extracting a simple semantic network from it. Since to date, no other system could do the same, we had to modify three other relational clustering approaches so that they could run on our large Web-scale dataset, and compared SNE to them. The three approaches are MRC [58], IRM [54], and Information-Theoretic Co-clustering (ITC) [28].

In order to make MRC and IRM feasible on large Web-scale datasets such as the one used in our experiments, we used the SNE clustering algorithm in place of the MRC and IRM search algorithms. We use MRC1 to denote an MRC model that is restricted to find a single clustering. For IRM, we also fixed the values of the CRP and Beta parameters and applied the CRP prior to cluster combinations instead of clusters.

The ITC model [28] clusters discrete data in a two-dimensional matrix along both dimensions simultaneously. It greedily searches for the hard clusterings that optimize the mutual information between the row and column clusters. The model has been shown to perform well on noisy and sparse data. ITC's top-down search algorithm has the flavor of K-means, and requires the number of row and column clusters to be specified in advance. At every step, ITC finds the best cluster for

Table 6.7: Semantic network comparison on gold standard. Object 1 and Object 2 respectively refer to the object symbols that appear as the first and second arguments of relations.

Model	Relation			Object 1			Object 2		
	Prec.	Recall	F1	Prec.	Recall	F1	Prec.	Recall	F1
SNE	0.452	0.187	0.265	0.460	0.061	0.108	0.558	0.062	0.112
MRC1	0.054	0.044	0.049	0.031	0.007	0.012	0.05	0.011	0.018
IRM	0.201	0.089	0.124	0.252	0.043	0.073	0.307	0.041	0.072
ITC	0.773	0.003	0.006	0.470	0.047	0.085	0.764	0.002	0.004

each row or column by iterating through all clusters. This is not tractable for large datasets like our Web dataset, which can contain many clusters; instead, we use the SNE clustering algorithm, but with mutual information in place of change in log posterior probability. We further extended ITC to three dimensions by optimizing the mutual information among the clusters of three dimensions. We follow Dhillon $et\ al$.'s [28] suggestion of using a BIC prior, but we applied it to the number of cluster combinations that contain at least one true ground atom, not the number of clusters.

DATASET
We compared the various models on a dataset of about 2.1 million triples[4] extracted in a Web crawl by TextRunner [3]. Each triple takes the form $r(x, y)$ where r is a relation symbol, and x and y are object symbols. Some example triples are: Named_after(Jupiter, Roman_god) and Upheld(Court, Ruling). There are 15,872 distinct r symbols, 700,781 distinct x symbols, and 665,378 distinct y symbols. Two characteristics of TextRunner's extractions are that they are sparse and noisy. To reduce the noise in the dataset, our search algorithm only considered symbols that appeared at least 25 times. This leaves 10,214 r symbols, 8942 x symbols, and 7995 y symbols. There are 2,065,045 triples that contain at least one symbol that appears at least 25 times. We make the closed-world assumption for all models (i.e., all triples not in the dataset are assumed false).

RESULTS
We evaluated the clusterings learned by each model against a manually created gold standard. The gold standard assigns 2688 r symbols, 2568 x symbols, and 3058 y symbols to 874, 511, and 700 non-unit clusters respectively. We measured the pairwise precision, recall and F1 of each model against the gold standard. Pairwise precision is the fraction of symbol pairs in learned clusters that appear in the same gold clusters. Pairwise recall is the fraction of symbol pairs in gold clusters that appear in the same learned clusters. F1 is the harmonic mean of precision and recall. Table 6.7 shows the results. We see that SNE performs significantly better than MRC1, IRM, and ITC.

[4]Publicly available at http://knight.cis.temple.edu/~yates/data/resolver_data.tar.gz

Table 6.8: Evaluation of semantic statements learned by SNE, IRM, and ITC.

Model	Total Statements	Number Correct	Fraction Correct
SNE	1241	965	0.778
IRM	487	426	0.874
ITC	310	259	0.835

We also ran MRC to find multiple clusterings. Since the gold standard only defines a single clustering, we cannot use it to evaluate the multiple clusterings. We provide a qualitative evaluation instead. MRC returns 23,151 leaves that contain non-unit clusters, and 99.8% of these only contain 3 or fewer clusters of size 2. In contrast, SNE finds many clusters of varying sizes. The poor performance of MRC in finding multiple clusterings is due to data sparsity. In each recursive call to MRC, it only receives a small subset of the relation and object symbols. Thus the data becomes sparser with each call and there is not enough signal to cluster the symbols.

We then evaluated SNE, IRM and ITC in terms of the semantic statements that they learned. A cluster combination that contains a true ground atom corresponds to a semantic statement. SNE, IRM and ITC respectively learned 1,464,965, 1,254,995 and 82,609 semantic statements. We manually inspected semantic statements containing 5 or more true ground atoms, and counted the number that were correct. Table 6.8 shows the results. Even though SNE's accuracy is smaller than IRM's and ITC's by 11% and 7% respectively, SNE more than compensates for the lower accuracy by learning 127% and 273% more correct statements, respectively. Figure 6.3 shows examples of correct semantic statements learned by SNE.

COMPARISON OF SNE WITH WORDNET

We also compared the object clusters that SNE learned with WordNet [34], a hand-built semantic lexicon for the English language. WordNet organizes 117,798 distinct nouns into a taxonomy of 82,115 concepts. There are 4883 first-argument and 5076 second-argument object symbols that appear at least 25 times in our dataset, and also in WordNet. We converted each node (synset) in WordNet's taxonomy into a cluster containing its original concepts and all its children concepts. We then matched each SNE cluster to the WordNet cluster that gave the best F1 score. We found that the scores were fairly good for object clusters of size seven or less, and that the object clusters correspond to fairly specific concepts in WordNet. We did not compare the relation clusters to WordNet's verbs because the overlap between the relation symbols and the verbs is too small.

FURTHER READING

For more details on the specific applications and experiments discussed here, see Lowd and Domingos [75] for collective classification; Richardson and Domingos [117] for link analysis in social

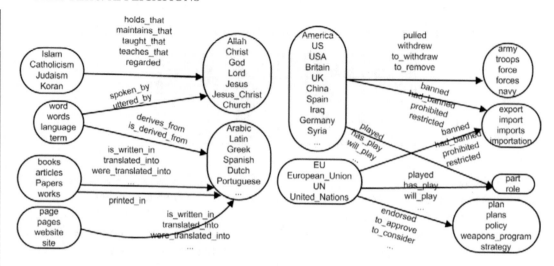

Figure 6.3: Fragments of a semantic network learned by SNE. Nodes are concept clusters, and the labels of links are relation clusters.

networks; Singla and Domingos [134] and Lowd and Domingos [75] for entity resolution; Poon and Domingos [106] for information extraction; Poon and Domingos [107] for coreference resolution; Wang and Domingos for robot mapping [148]; Kok and Domingos [58] for link-based clustering; and Kok and Domingos [59] for semantic network extraction from text.

The number of applications of Markov logic is quickly growing. Other applications include the DARPA CALO personal assistant [29], semantic role labeling in NLP [120], extracting genic interactions from Medline abstracts [119], predicting protein β-partners [70], extracting ontologies from Wikipedia [154], discovering social relationships in photo collections [139], using the Web to answer queries [130], and more.

CHAPTER 7

Conclusion

Progress in an area of computer science picks up when the right *interface layer* is defined: an abstract language that separates the infrastructure from the applications, enabling each piece of the infrastructure to communicate with all the applications without modification, and vice-versa. Markov logic is an attempt to provide such an interface layer for artificial intelligence.

Markov logic combines first-order logic and probabilistic graphical models. A knowledge base in Markov logic is simply a set of weighted first-order formulas, viewed as templates for features of a Markov network: given a set of constants representing objects in the domain of interest, the Markov network has one Boolean variable for each possible grounding of each predicate with those constants (e.g., Friends(Anna, Bob), and one Boolean feature for each grounding of each formula, with the corresponding weight (e.g., Smokes(Anna) ∧ Friends(Anna, Bob) ⇒ Smokes(Bob)). First-order logic is the infinite-weight limit of Markov logic.

Most widely used graphical models can be specified very compactly in Markov logic (e.g., two formulas containing a total of three predicates for logistic regression, three formulas for a hidden Markov model, etc.). In turn, this makes it easy to define large, complex models involving combinations of these components. For example, a complete information extraction system (performing joint segmentation and entity resolution) can be specified with just a few formulas, and a state-of-the-art one with a few dozen.

Several implementations of Markov logic exist, of which the most developed is Alchemy [60]. Alchemy includes a series of highly efficient inference and learning algorithms for Markov logic, including:

- A weighted satisfiability solver for MAP inference. This and other algorithms use lazy inference techniques to minimize the number of ground predicates that are actually created.

- MC-SAT, an MCMC algorithm that uses a satisfiability solver to find modes, for computing conditional probabilities.

- Lifted belief propagation, which allows probabilistic inference to be carried out for whole sets of objects at once, without grounding predicates, like resolution-based theorem proving in Prolog.

- Gradient descent, conjugate gradient and quasi-Newton methods for generative and discriminative weight learning.

- Inductive logic programming algorithms for learning new formulas from data and refining manually specified ones.

These algorithms make it possible to routinely perform learning and inference on graphical models with millions of variables and billions of features. State-of-the-art solutions to a wide variety of problems have been developed using Markov logic, including:

- The first system for unsupervised coreference resolution in NLP that outperforms supervised ones on standard benchmarks.

- The first robot mapping system that performs joint labeling of range finder segments, clustering of segments into walls, and regression to estimate the locations of walls from their segments.

- The winning entry in the LLL-05 competition on extracting gene interactions from Medline abstracts.

- An inference engine for the DARPA CALO personal assistant.

- The most accurate system to date for extracting ontologies from Wikipedia.

- The first unsupervised system for extracting semantic networks from Web crawls.

- A state-of-the-art system for predicting protein β-partners.

- A probabilistic version of (parts of) Cyc, the world's largest knowledge base.

Markov logic makes the prospect of "programming by machine learning" a realistic one. With Alchemy, the programmer's work is reduced to specifying the approximate structure of the program; the rest is filled in by learning from data, and the resulting uncertainties are handled by probabilistic inference. With discriminative learning, Markov logic learns the conditional distribution of the program outputs given its inputs, and the most likely program is run by performing MAP inference.

Even without learning, weighted logic makes it possible to write knowledge bases that are much more compact than the corresponding ones in purely logical languages. It also relaxes the requirements for perfect correctness, consistency and completeness that make writing large KBs and integrating contributions from many sources a very costly (or even infeasible) enterprise.

Markov logic and Alchemy are not yet at the point where they can be easily used by developers or researchers without expertise in machine learning, automated reasoning, or probabilistic inference. The chief reason for this is that the underlying inference and learning algorithms still require tuning, parameter choices, trial and error, etc., before they produce the desired results. Markov logic successfully integrates learning, logic and probability, but it inherits the state of the art in each of these fields, and in each of them there is still much research to be done before "push-button" algorithms are available.

However, even with this state of art, Markov logic already allows us to attempt much more ambitious applications than before. For example, we are currently developing a complete natural language processing system, using fully joint inference across all stages, and ultimately consisting of probably no more than hundreds of (second-order) Markov logic formulas.

It is our hope that researchers in machine learning, probabilistic inference, and knowledge representation and reasoning will find it worthwhile to provide their latest algorithms as plug-ins to Alchemy using the corresponding APIs, thus making them available to all applications that use MLNs as the interface layer. Conversely, we hope that researchers in natural language, planning, vision and robotics, multi-agent systems, etc., will find Markov logic a useful modeling tool, and contribute the MLNs they develop to the Alchemy repository, making them available for other researchers to build on. Thus Markov logic will help speed progress toward the ultimate goal of reaching and surpassing human-level intelligence.

APPENDIX A

The Alchemy System

Alchemy is an open-source system that aims to make many of the models and algorithms discussed in this book easily accessible to researchers and practitioners. Source code, documentation, tutorials, MLNs, datasets, and publications are available at the Alchemy Web site, `http://alchemy.cs.washington.edu`. In this appendix, we provide a brief introduction to the Alchemy system.

Alchemy includes algorithms for learning and inference with Markov logic networks and hybrid Markov logic networks. This includes generative and discriminative weight learning, top-down structure learning,[1] MAP/MPE inference, and probabilistic inference. Second-order MLNs, recursive random fields, and other extensions are not supported at this time.

Alchemy was developed under Linux, but also works under Windows using Cygwin. (To run Alchemy on a Mac, see additional notes online.) To install Alchemy, first download the source code distribution from the Web site. Unpacking the downloaded file will create an `alchemy` directory as well as various subdirectories, including `src`. To build Alchemy, run `make depend; make` from the `alchemy/src` directory. In addition to the `make` utility, compiling Alchemy also requires `bison`, `flex`, `g++`, and `perl`.

A.1 INPUT FILES

Predicates and functions are declared and first-order formulas are specified in `.mln` files. The first appearance of a predicate or function in a `.mln` file is taken to be its declaration. You can express arbitrary first-order formulas in a `.mln` file. A variable in a formula must begin with a lowercase character, and a constant with an uppercase one in a formula. A formula can be preceded by a weight or terminated by a period, but not both. A period signifies that a formula is "hard" (i.e., worlds that violate it should have zero probability). Types and constants are also declared in `.mln` files. As in C++ and Java, comments can be included with `//` and `/* */`. Blank lines are ignored.

Table A.1 contains an example MLN to model friendship ties between people, their smoking habits and the causality of cancer. The MLN begins by declaring the predicates `Friends`, `Smokes`, and `Cancer`, each taking one or two arguments of type `person`. If we wished, we could declare the person type explicitly by listing its constants, e.g., `person = {Anna, Bob, Chris}`. Without a type declaration, the constants will automatically be inferred from the database. The two formulas link smoking to cancer and the smoking habits of friends. When learning, weights may be omitted or used to specify the prior means.

---

[1]Bottom-up structure learning has been implemented as an extension to Alchemy, available separately from `http://www.cs.utexas.edu/~ml/mlns/`.

Ground atoms are defined in .db (database) files. Ground atoms preceded by ! (e.g., !Friends(Anna,Bob)) are false, by ? are unknown, and by neither are true. If the closed world assumption is made for a predicate, its ground atoms that are not defined in a .db file are false, while if the open world assumption is made, its undefined ground atoms are unknown. By default, all predicates are closed world during learning and only the specified query predicates are open world during inference. Table A.2 shows an example database for the "friends and smokers" social network domain.

Table A.1: smoking.mln, example MLN for friends and smokers domain.

```
//predicate declarations
Friends(person, person)
Smokes(person)
Cancer(person)


// If you smoke, you get cancer
1.5 Smokes(x) => Cancer(x)


// Friends have similar smoking habits
0.8 Friends(x, y) => (Smokes(x) <=> Smokes(y))
```

Table A.2: smoking-train.db, an example training database for smoking.mln.

```
Friends(Anna, Bob)
Friends(Bob, Anna)
Friends(Anna, Edward)
Friends(Edward, Anna)
Friends(Bob, Chris)
Friends(Chris, Bob)
Friends(Chris, Daniel)
Friends(Daniel, Chris)
Smokes(Anna)
Smokes(Bob)
Smokes(Edward)
Cancer(Anna)
Cancer(Edward)
```

Although not included in this example, Alchemy also supports user-defined functions. Functions are declared in a .mln file, like predicates but with a return type, e.g., person MotherOf(person). Function values are specified in a .db file. Each line contains exactly one definition of the form <*returnvalue*> = <functionname>(<*constant1*>, ..., <*constantn*>), e.g., Anna = MotherOf(Bob). The mappings given are assumed to be the only ones present in the domain. It is also possible to link in externally defined functions and predicates.

Alchemy supports the + notation and ! notation described in Section 6.1. See the online user manual for more information on Alchemy syntax and other features.

A.2   INFERENCE

To perform inference, run the infer executable, e.g.:

```
ALCHDIR/bin/infer -i smoking.mln -e smoking.db -r smoking.results
        -q Smokes -ms -maxSteps 20000
```

-i specifies the input .mln file. All formulas in the input file must be preceded by a weight or terminated by a period (but not both). -e specifies the evidence .db file; a comma-separated list can be used to specify more than one .db file. -r specifies the output file, which contains the inference results.

-q specifies the query predicates. You can specify more than one query predicate, and restrict the query to particular groundings, e.g., -q Smokes(Anna),Smokes(Bob). (Depending on the shell you are using, you may have to enclose the query predicates in quotes because of the presence of parentheses.) You can also use the -f option to specify a file containing the query ground atoms you are interested in (same format as a .db file without false and unknown atoms). Query atoms in the evidence database are assumed to be known and will not be inferred.

The output from the example inference command looks something like this:

```
Smokes(Chris) 0.238926
Smokes(Daniel) 0.141286
```

The probability of Anna, Bob and Edward smoking is not computed since their smoking status is already known in the database.

Alchemy supports two basic types of inference: computing conditional probabilities (using MCMC or lifted belief propagation) and finding the most probable state of the world (MAP/MPE inference). The current implementation includes three MCMC algorithms: Gibbs sampling (option -p), MC-SAT (option -ms) and simulated tempering (option -simtp). -maxSteps is used to specify the maximum number of steps in the MCMC algorithm. Belief propagation is specified with option -bp. When MCMC inference or belief propagation is run, the probabilities that the query atoms are true are written to the output file specified. To use MAP inference instead, specify either the -m or -a option. The former only returns the true ground atoms, while the latter returns both true and false ones. All inference options except belief propagation support the -lazy flag to avoid grounding the entire domain (see Section 3.3). Belief propagation supports lifted inference (-lifted) instead (see Section 3.4).

Type ALCHDIR/bin/infer without any parameters to view all options.

A.3   WEIGHT LEARNING

To learn the weights of formulas, run the learnwts executable, e.g.:

```
ALCHDIR/bin/learnwts -d -i smoking.mln -o smoking-out.mln
        -t smoking-train.db -ne Smokes,Cancer
```

where ALCHDIR is the directory Alchemy is installed in. -d specifies discriminative learning; for generative learning (pseudo-log-likelihood), use -g instead. -i and -o specify the input and output .mln files. -t specifies the .db file that is to be used by weight learning. You can specify multiple .db or .mln files in a comma separated list (e.g., -t smoking1.db,smoking2.db). By default, constants are shared across databases. To specify instead that the constants in each .db file belong to a separate database, use the -multipleDatabases option. The -ne option specifies the non-evidence predicates, those that will be unknown and inferred at inference time.

During weight learning, each formula is converted to conjunctive normal form (CNF), and a weight is learned for each of its clauses. If the formulas in the input MLN have weights, these can be used as the means of Gaussian priors on the weights to be learned, with the standard deviation specified by -priorStdDev <double>. A global prior mean can be specified by -priorMean <double>; the default is 0.

After weight learning, the output .mln file contains the weights of the original formulas (commented out), as well as those of their derived clauses. By default, unit clauses for all predicates are added to the MLN during weight learning. (You can change this with the -noAddUnitClauses option.)

Table A.3 shows the result of performing discriminative weight learning with the example MLN and database from Tables A.1 and A.2. Note that the second formula has been split into two clauses to capture both directions of the equivalence. The unit clause Friends(a1, a2) receives a weight of zero because it was not specified as a non-evidence predicate, and is therefore assumed to be known.

You can view all the options by typing ALCHDIR/bin/learnwts without any parameters. One of the most important is -dNumIters <integer>, which sets the maximum number of iterations for discriminative weight learning. Inference options (including -lazy) can also be used with learnwts when embedded in the -infer option, e.g.: -infer "-ms -maxSteps 100 -lazy".

A.4   STRUCTURE LEARNING

To learn the structure (clauses and weights) of an MLN generatively, you use the learnstruct executable, e.g.:

Table A.3: `smoking-out.mln`, learned MLN for friends and smokers domain.

```
//predicate declarations
Cancer(person)
Friends(person,person)
Smokes(person)

//function declarations

// 1.51903   Smokes(x) => Cancer(x)
1.51903   !Smokes(a1) v Cancer(a1)

// 0.994841  Friends(x,y) => (Smokes(x) <=> Smokes(y))
0.49742   !Friends(a1,a2) v Smokes(a1) v !Smokes(a2)
0.49742   !Friends(a1,a2) v Smokes(a2) v !Smokes(a1)

// 0         Friends(a1,a2)
0         Friends(a1,a2)

// 0.86298   Smokes(a1)
0.86298   Smokes(a1)

// -1.0495   Cancer(a1)
-1.0495   Cancer(a1)
```

```
ALCHDIR/bin/learnstruct -i smoking.mln -o smoking-out.mln
        -t smoking.db -penalty 0.5
```

`learnstruct` uses beam search to find new clauses to add to an MLN. It can start from both empty and non-empty MLNs. Its options are similar to those of `learnwts`. In addition, it has options for controlling techniques that speed up the search. You can also restrict the types of clauses created during structure learning (see the developer's manual online). Type `ALCHDIR/bin/learnstruct` without any parameters to view all options.

FURTHER READING

For full documentation on Alchemy, see `http://alchemy.cs.washington.edu`. Many more details about syntax and options are available in the user's guide; information about the internal workings of Alchemy is in the developer's guide; and a gentle tutorial (including MLN and database files) covers much of Alchemy's functionality through examples. Additional support is available by email: `alchemy@cs.washington.edu`. The `alchemy-discuss@cs.washington.edu` mailing list is dedicated to discussing topics related to Alchemy.

Bibliography

[1] F. Bacchus. *Representing and Reasoning with Probabilistic Knowledge*. MIT Press, Cambridge, MA, 1990.

[2] F. Bacchus, A. J. Grove, J. Y. Halpern, and D. Koller. From statistical knowledge bases to degrees of belief. *Artificial Intelligence*, 87:75–143, 1996. DOI: 10.1016/S0004-3702(96)00003-3

[3] M. Banko, M. J. Cafarella, S. Soderland, M. Broadhead, and O. Etzioni. Open information extraction from the web. In *Proceedings of the Twentieth International Joint Conference on Artificial Intelligence*, pages 2670–2676, Hyderabad, India, 2007. AAAI Press. DOI: 10.1145/1409360.1409378

[4] J. O. Berger. *Statistical Decision Theory and Bayesian Analysis*. Springer, New York, NY, 1985.

[5] T. Berners-Lee, J. Hendler, and O. Lassila. The Semantic Web. *Scientific American*, 284(5):34–43, 2001.

[6] J. Besag. Statistical analysis of non-lattice data. *The Statistician*, 24:179–195, 1975. DOI: 10.2307/2987782

[7] J. Besag. On the statistical analysis of dirty pictures. *Journal of the Royal Statistical Society, Series B*, 48:259–302, 1986. DOI:

[8] M. Biba, S. Ferilli, and F. Esposito. Discriminative structure learning of Markov logic networks. In *Proceedings of Eighteenth International Conference on Inductive Logic Programming*, pages 59–76, Prague, Czech Republic, 2008. Springer. DOI: 10.1007/978-3-540-85928-4_9

[9] P. Billingsley. *Probability and Measure*. John Wiley and Sons, New York, 1995.

[10] C. Boutilier, R. Reiter, and B. Price. Symbolic dynamic programming for first-order MDPs. In *Proceedings of the Seventeenth International Joint Conference on Artificial Intelligence*, pages 690–697, Seattle, WA, 2001. Morgan Kaufmann.

[11] R. Braz, E. Amir, and D. Roth. Lifted first-order probabilistic inference. In *Proceedings of the Nineteenth International Joint Conference on Artificial Intelligence*, pages 1319–1325, Edinburgh, UK, 2005. Morgan Kaufmann.

[12] R. Braz, E. Amir, and D. Roth. MPE and partial inversion in lifted probabilistic variable elimination. In *Proceedings of the Twenty-First National Conference on Artificial Intelligence*, pages 1123–1130, Boston, MA, 2006. AAAI Press.

[13] F. Bromberg, D. Margaritis, and V. Honavar. Efficient Markov network structure discovery from independence tests. In *SIAM International Conference on Data Mining*, pages 141–152, Bethesda, MD, 2006.

[14] W. Buntine. Operations for learning with graphical models. *Journal of Artificial Intelligence Research*, 2:159–225, 1994.

[15] M. Collins. Discriminative training methods for hidden Markov models: Theory and experiments with perceptron algorithms. In *Proceedings of the 2002 Conference on Empirical Methods in Natural Language Processing*, pages 1–8, Philadelphia, PA, 2002. ACL. DOI: 10.3115/1118693.1118694

[16] M. Craven and S. Slattery. Relational learning with statistical predicate invention: Better models for hypertext. *Machine Learning*, 43(1/2):97–119, 2001. DOI: 10.1023/A:1007676901476

[17] J. Cussens. Loglinear models for first-order probabilistic reasoning. In *Proceedings of the Fifteenth Conference on Uncertainty in Artificial Intelligence*, pages 126–133, Stockholm, Sweden, 1999. Morgan Kaufmann.

[18] J. Cussens. Individuals, relations and structures in probabilistic models. In *Proceedings of the IJCAI-2003 Workshop on Learning Statistical Models from Relational Data*, pages 32–36, Acapulco, Mexico, 2003. IJCAII.

[19] P. Damien, J. Wakefield, and S. Walker. Gibbs sampling for Bayesian non-conjugate and hierarchical models by auxiliary variables. *Journal of the Royal Statistical Society, Series B*, 61:331–344, 1999. DOI: 10.1111/1467-9868.00179

[20] J. Davis and P. Domingos. Deep transfer via second-order Markov logic. In *Proceedings of the Twenty-Sixth International Conference on Machine Learning*, Montréal, Canada, 2009. ACM Press.

[21] J. Davis, I. Ong, J. Struyf, E. Burnside, D. Page, and V. S. Costa. Change of representation for statistical relational learning. In *Proceedings of the Twentieth International Joint Conference on Artificial Intelligence*, pages 2719–2726, Hyderabad, India, 2007. AAAI Press.

[22] L. De Raedt and L. Dehaspe. Clausal discovery. *Machine Learning*, 26:99–146, 1997. DOI: 10.1023/A:1007361123060

[23] L. Dehaspe. Maximum entropy modeling with clausal constraints. In *Proceedings of the Seventh International Workshop on Inductive Logic Programming*, pages 109–125, Prague, Czech Republic, 1997. Springer.

[24] S. Della Pietra, V. Della Pietra, and J. Lafferty. Inducing features of random fields. *IEEE Transactions on Pattern Analysis and Machine Intelligence*, 19:380–392, 1997. DOI: 10.1109/34.588021

[25] A. P. Dempster, N. M. Laird, and D. B. Rubin. Maximum likelihood from incomplete data via the EM algorithm. *Journal of the Royal Statistical Society, Series B*, 39:1–38, 1977. DOI: 10.2307/2984875

[26] W. Denham. *The detection of patterns in Alyawarra nonverbal behavior*. PhD thesis, Department of Anthropology, University of Washington, Seattle, WA, 1973.

[27] P. Denis and J. Baldridge. Joint determination of anaphoricity and coreference resolution using integer programming. In *Proceedings of NAACL HLT 2007*, pages 236–243, Rochester, New York, 2007. ACL.

[28] I. S. Dhillon, S. Mallela, and D. S. Modha. Information-theoretic co-clustering. In *Proceedings of the Ninth ACM SIGKDD International Conference on Knowledge Discovery and Data Mining*, pages 89–98, Washington, DC, 2003. ACM Press. DOI: 10.1145/956750.956764

[29] T. Dietterich and X. Bao. Integrating multiple learning components through Markov logic. In *Proceedings of the Twenty-Third National Conference on Artificial Intelligence*, pages 622–627, Chicago, IL, 2008. AAAI Press.

[30] P. Domingos and M. Pazzani. On the optimality of the simple Bayesian classifier under zero-one loss. *Machine Learning*, 29:103–130, 1997. DOI: 10.1023/A:1007413511361

[31] K. Driessens, J. Ramon, and T. Croonenborghs. Transfer learning for reinforcement learning through goal and policy parametrization. In *Proceedings of the ICML 2006 Workshop on Structural Knowledge Transfer for Machine Learning*, 2006.

[32] S. Džeroski and L. De Raedt. Relational reinforcement learning. In *Proceedings of the Fifteenth International Conference on Machine Learning*, pages 136–143, Madison, WI, 1998. Morgan Kaufmann. DOI: 10.1023/A:1007694015589

[33] S. Džeroski and N. Lavrač, editors. *Relational Data mining*. Springer, Berlin, Germany, 2001.

[34] C. Fellbaum, editor. *WordNet: An Electronic Lexical Database*. MIT Press, Cambridge, MA, 1998.

[35] R. Fletcher. *Practical Methods of Optimization*. Wiley-Interscience, New York, NY, second edition, 1987.

[36] N. Friedman, L. Getoor, D. Koller, and A. Pfeffer. Learning probabilistic relational models. In *Proceedings of the Sixteenth International Joint Conference on Artificial Intelligence*, pages 1300–1307, Stockholm, Sweden, 1999. Morgan Kaufmann.

[37] M. R. Genesereth and N. J. Nilsson. *Logical Foundations of Artificial Intelligence*. Morgan Kaufmann, San Mateo, CA, 1987.

[38] H Georgii. *Gibbs Measures and Phase Transitions*. Walter de Gruyter, Berlin, 1988.

[39] L. Getoor and B. Taskar, editors. *Introduction to Statistical Relational Learning*. MIT Press, Cambridge, MA, 2007.

[40] W. R. Gilks, S. Richardson, and D. J. Spiegelhalter, editors. *Markov Chain Monte Carlo in Practice*. Chapman and Hall, London, UK, 1996.

[41] T. Grenager, D. Klein, and C. D. Manning. Unsupervised learning of field segmentation models for information extraction. In *Proceedings of the Forty-Third Annual Meeting on Association for Computational Linguistics*, pages 371–378, Ann Arbor, Michigan, 2005. Association for Computational Linguistics. DOI: 10.3115/1219840.1219886

[42] A. Haghighi and D. Klein. Unsupervised coreference resolution in a nonparametric Bayesian model. In *Proceedings of the 45th Annual Meeting of the Association for Computational Linguistics*, pages 848–855, Prague, Czech Republic, 2007. ACL.

[43] J. Halpern. An analysis of first-order logics of probability. *Artificial Intelligence*, 46:311–350, 1990. DOI: 10.1016/0004-3702(90)90019-V

[44] D. Heckerman, D. M. Chickering, C. Meek, R. Rounthwaite, and C. Kadie. Dependency networks for inference, collaborative filtering, and data visualization. *Journal of Machine Learning Research*, 1:49–75, 2000. DOI: 10.1162/153244301753344614

[45] D. Heckerman, D. Geiger, and D. M. Chickering. Learning Bayesian networks: The combination of knowledge and statistical data. *Machine Learning*, 20:197–243, 1995. DOI: 10.1007/BF00994016

[46] D. Heckerman, C. Meek, and D. Koller. Probabilistic entity-relationship models, PRMs, and plate models. In *Proceedings of the ICML-2004 Workshop on Statistical Relational Learning and its Connections to Other Fields*, pages 55–60, Banff, Canada, 2004. IMLS.

[47] G. E. Hinton. Training products of experts by minimizing contrastive divergence. *Neural Computation*, 14(8):1771–1800, 2002. DOI: 10.1162/089976602760128018

[48] R. A. Howard and J. E. Matheson. Influence diagrams. *Decision Analysis*, 2(3):127–143, 2005. DOI: 10.1287/deca.1050.0020

[49] G. Hulten and P. Domingos. Mining complex models from arbitrarily large databases in constant time. In *Proceedings of the Eighth ACM SIGKDD International Conference on Knowledge Discovery and Data Mining*, pages 525–531, Edmonton, Canada, 2002. ACM Press. DOI: 10.1145/775047.775124

[50] T. Huynh and R. Mooney. Discriminative structure and parameter learning for Markov logic networks. In *Proceedings of the Twenty-Fifth International Conference on Machine Learning*, pages 416–423, Helsinki, Finland, 2008. ACM Press. DOI: 10.1145/1390156.1390209

[51] A. Jaimovich, O. Meshi, and N. Friedman. Template based inference in symmetric relational Markov random fields. In *Proceedings of the Twenty-Third Conference on Uncertainty in Artificial Intelligence*, pages 191–199, Vancouver, Canada, 2007. AUAI Press.

[52] H. Kautz and B. Selman. Planning as satisfiability. In *Proceedings of the Tenth European Conference on Artificial Intelligence*, pages 359–363, New York, NY, 1992. Wiley.

[53] H. Kautz, B. Selman, and Y. Jiang. A general stochastic approach to solving problems with hard and soft constraints. In D. Gu, J. Du, and P. Pardalos, editors, *The Satisfiability Problem: Theory and Applications*, pages 573–586. American Mathematical Society, New York, NY, 1997.

[54] C. Kemp, J. B. Tenenbaum, T. L. Griffiths, T. Yamada, and N Ueda. Learning systems of concepts with an infinite relational model. In *Proceedings of the Twenty-First National Conference on Artificial Intelligence*, pages 381–388, Boston, MA, 2006. AAAI Press.

[55] K. Kersting and L. De Raedt. Towards combining inductive logic programming with Bayesian networks. In *Proceedings of the Eleventh International Conference on Inductive Logic Programming*, pages 118–131, Strasbourg, France, 2001. Springer. DOI: 10.1007/3-540-44797-0

[56] D. Klein and C. Manning. Accurate unlexicalized parsing. In *Proceedings of the 41st Annual Meeting of the Association for Computational Linguistics*, pages 423–430, Sapporo, Japan, 2003. ACL. DOI: 10.3115/1075096.1075150

[57] S. Kok and P. Domingos. Learning the structure of Markov logic networks. In *Proceedings of the Twenty-Second International Conference on Machine Learning*, pages 441–448, Bonn, Germany, 2005. ACM Press. DOI: 10.1145/1102351.1102407

[58] S. Kok and P. Domingos. Statistical predicate invention. In *Proceedings of the Twenty-Fourth International Conference on Machine Learning*, pages 433–440, Corvallis, OR, 2007. ACM Press. DOI: 10.1145/1273496.1273551

[59] S. Kok and P. Domingos. Extracting semantic networks from text via relational clustering. In *Proceedings of the Nineteenth European Conference on Machine Learning*, pages 624–639, Antwerp, Belgium, 2008. Springer. DOI: 10.1007/978-3-540-87479-9_59

[60] S. Kok, M. Sumner, M. Richardson, P. Singla, H. Poon, D. Lowd, and P. Domingos. The Alchemy system for statistical relational AI. Technical report, Department of Computer Science and Engineering, University of Washington, Seattle, WA, 2007. http://alchemy.cs.washington.edu.

[61] D. Koller and N. Friedman. *Structured Probabilistic Models: Principles and Techniques*. MIT Press, Cambridge, MA, 2009.

[62] F. R. Kschischang, B. J. Frey, and H.-A. Loeliger. Factor graphs and the sum-product algorithm. *IEEE Transactions on Information Theory*, 47:498–519, 2001. DOI: 10.1109/18.910572

[63] N. Kushmerick. Wrapper induction: Efficiency and expressiveness. *Artificial Intelligence*, 118(1-2):15–68, 2000. DOI: 10.1016/S0004-3702(99)00100-9

[64] J. Laffar and J.-L. Lassez. Constraint logic programming. In *Proceedings of the Fourteenth ACM Conference on Principles of Programming Languages*, pages 111–119, Munich, Germany, 1987. ACM Press. DOI: 10.1145/41625.41635

[65] N. Lavrač and S. Džeroski. *Inductive Logic Programming: Techniques and Applications*. Ellis Horwood, Chichester, UK, 1994.

[66] S. Lawrence, K. Bollacker, and C. L. Giles. Autonomous citation matching. In *Proceedings of the Third International Conference on Autonomous Agents*, pages 392–393, New York, 1999. ACM Press. DOI: 10.1145/301136.301255

[67] S.-I. Lee, V. Chatalbashev, D. Vickrey, and D. Koller. Learning a meta-level prior for feature relevance from multiple related tasks. In *Proceedings of the Twenty-Fourth International Conference on Machine Learning*, pages 489–496, Corvallis, OR, 2007. DOI: 10.1145/1273496.1273558

[68] P. Liang and M. I. Jordan. An asymptotic analysis of generative, discriminative, and pseudolikelihood estimators. In *Proceedings of the Twenty-Fifth International Conference on Machine Learning*, pages 584–591, Helsinki, Finland, 2008. ACM Press. DOI: 10.1145/1390156.1390230

[69] B. Limketkai, L. Liao, and D. Fox. Relational object maps for mobile robots. In *Proceedings of the Nineteenth International Joint Conference on Artificial Intelligence*, pages 1471–1476, Edinburgh, UK, 2005. Morgan Kaufmann.

[70] M. Lippi and P. Frasconi. Markov logic improves protein β-partners prediction. In *Proceedings of the Sixth International Workshop on Mining and Learning with Graphs*, Helsinki, Finland, 2008.

[71] D. C. Liu and J. Nocedal. On the limited memory BFGS method for large scale optimization. *Mathematical Programming*, 45(3):503–528, 1989. DOI: 10.1007/BF01589116

[72] J. W. Lloyd. *Foundations of Logic Programming*. Springer, Berlin, Germany, 1987.

[73] E. Lloyd-Richardson, A. Kazura, C. Stanton, R. Niaura, and G. Papandonatos. Differentiating stages of smoking intensity among adolescents: Stage-specific psychological and social influences. *Journal of Consulting and Clinical Psychology*, 70(4):998–1009, 2002. DOI: 10.1037/0022-006X.70.4.998

[74] B. Long, Z. M. Zhang, X. Wu, and P. S. Yu. Spectral clustering for multi-type relational data. In *Proceedings of the Twenty-Third International Conference on Machine Learning*, pages 585–592, Pittsburgh, PA, 2006. ACM Press. DOI: 10.1145/1143844.1143918

[75] D. Lowd and P. Domingos. Efficient weight learning for Markov logic networks. In *Proceedings of the Eleventh European Conference on Principles and Practice of Knowledge Discovery in Databases*, pages 200–211, Warsaw, Poland, 2007. Springer. DOI: 10.1007/978-3-540-74976-9_21

[76] D. Lowd and P. Domingos. Recursive random fields. In *Proceedings of the Twentieth International Joint Conference on Artificial Intelligence*, pages 950–955, Hyderabad, India, 2007. AAAI Press.

[77] X. Luo, A. Ittycheriah, H. Jing, N. Kambhatla, and S. Roukos. A mention-synchronous coreference resolution algorithm based on the Bell tree. In *Proceedings of the 42nd Annual Meeting of the Association for Computational Linguistics*, pages 135–142, Barcelona, Spain, 2004. ACL. DOI: 10.3115/1218955.1218973

[78] A. McCallum. Efficiently inducing features of conditional random fields. In *Proceedings of the Nineteenth Conference on Uncertainty in Artificial Intelligence*, Acapulco, Mexico, 2003. Morgan Kaufmann.

[79] A. McCallum, R. Rosenfeld, T. Mitchell, and A. Y. Ng. Improving text classification by shrinkage in a hierarchy of classes. In *Proceedings of the Fifteenth International Conference on Machine Learning*, pages 359–367, Madison, WI, 1998. Morgan Kaufmann.

[80] A. McCallum and B. Wellner. Conditional models of identity uncertainty with application to noun coreference. In *Advances in Neural Information Processing Systems 17*, pages 905–912, Cambridge, MA, 2005. MIT Press.

[81] A. T. McCray. An upper level ontology for the biomedical domain. *Comparative and Functional Genomics*, 4:80–84, 2003. DOI: 10.1002/cfg.255

[82] L. Mihalkova, T. Huynh, and R. J. Mooney. Mapping and revising Markov logic networks for transfer learning. In *Proceedings of the Twenty-Second National Conference on Artificial Intelligence*, pages 608–614, Vancouver, Canada, 2007. AAAI Press.

[83] L. Mihalkova and R. Mooney. Bottom-up learning of Markov logic network structure. In *Proceedings of the Twenty-Fourth International Conference on Machine Learning*, pages 625–632, Corvallis, OR, 2007. ACM Press. DOI: 10.1145/1273496.1273575

[84] B. Milch, B. Marthi, S. Russell, D. Sontag, D. L. Ong, and A. Kolobov. BLOG: Probabilistic models with unknown objects. In *Proceedings of the Nineteenth International Joint Conference on Artificial Intelligence*, pages 1352–1359, Edinburgh, UK, 2005. Morgan Kaufmann.

[85] B. Milch, L. S. Zettlemoyer, K. Kersting, M. Haimes, and L. P. Kaelbling. Lifted probabilistic inference with counting formulas. In *Proceedings of the Twenty-Third National Conference on Artificial Intelligence*, pages 1062–1068, Chicago, IL, 2008. AAAI Press.

[86] M. Møller. A scaled conjugate gradient algorithm for fast supervised learning. *Neural Networks*, 6:525–533, 1993. DOI: 10.1016/S0893-6080(05)80056-5

[87] S. Muggleton. Stochastic logic programs. In L. De Raedt, editor, *Advances in Inductive Logic Programming*, pages 254–264. IOS Press, Amsterdam, Netherlands, 1996.

[88] S. Muggleton and C. Feng. Efficient induction of logic programs. In S. Muggleton, editor, *Inductive Logic Programming*, pages 281–298. Morgan Kaufmann, 1992.

[89] A. Nath and P. Domingos. A language for relational decision theory. In *Proceedings of the International Workshop on Statistical Relational Learning*, Leuven, Belgium, 2009.

[90] J. Neville and D. Jensen. Collective classification with relational dependency networks. In S. Džeroski, L. De Raedt, and S. Wrobel, editors, *Proceedings of the Second International Workshop on Multi-Relational Data Mining*, pages 77–91, Washington, DC, 2003. ACM Press.

[91] J. Neville and D. Jensen. Leveraging relational autocorrelation with latent group models. In *Proceedings of the Fifth IEEE International Conference on Data Mining*, pages 49–55, New Orleans, LA, 2005. IEEE Computer Society Press. DOI: 10.1109/ICDM.2005.89

[92] V. Ng. Machine learning for coreference resolution: From local classification to global ranking. In *Proceedings of the 43th Annual Meeting of the Association for Computational Linguistics*, pages 157–164, Ann Arbor, MI, 2005. ACL. DOI: 10.3115/1219840.1219860

[93] L. Ngo and P. Haddawy. Answering queries from context-sensitive probabilistic knowledge bases. *Theoretical Computer Science*, 171:147–177, 1997. DOI: 10.1016/S0304-3975(96)00128-4

[94] N. Nilsson. Probabilistic logic. *Artificial Intelligence*, 28:71–87, 1986. DOI: 10.1016/0004-3702(86)90031-7

[95] J. Nocedal and S. Wright. *Numerical Optimization*. Springer, New York, NY, 2006.

[96] D. N. Osherson, J. Stern, O. Wilkie, M. Stob., and E. E. Smith. Default probability. *Cognitive Science*, 15:251–269, 1991. DOI: 10.1016/0364-0213(91)80007-R

[97] M. Paskin. Maximum entropy probabilistic logic. Technical Report UCB/CSD-01-1161, Computer Science Division, University of California, Berkeley, CA, 2002.

[98] H. Pasula, B. Marthi, B. Milch, S. Russell, and I. Shpitser. Identity uncertainty and citation matching. In *Advances in Neural Information Processing Systems 14*, Cambridge, MA, 2002. MIT Press.

[99] J. Pearl. *Probabilistic Reasoning in Intelligent Systems: Networks of Plausible Inference.* Morgan Kaufmann, San Francisco, CA, 1988.

[100] B. Pearlmutter. Fast exact multiplication by the Hessian. *Neural Computation*, 6(1):147–160, 1994. DOI: 10.1162/neco.1994.6.1.147

[101] A. Pfeffer, D. Koller, B. Milch, and K. T. Takusagawa. Spook: A system for probabilistic object-oriented knowledge representation. In *Proceedings of the Fifteenth Conference on Uncertainty in Artificial Intelligence*, pages 541–550, Stockholm, Sweden, 1999. Morgan Kaufmann.

[102] J. Pitman. Combinatorial stochastic processes. Technical Report 621, Department of Statistics, University of California at Berkeley, Berkeley, CA, 2002.

[103] D. Poole. Probabilistic Horn abduction and Bayesian networks. *Artificial Intelligence*, 64:81–129, 1993. DOI: 10.1016/0004-3702(93)90061-F

[104] D. Poole. First-order probabilistic inference. In *Proceedings of the Eighteenth International Joint Conference on Artificial Intelligence*, pages 985–991, Acapulco, Mexico, 2003. Morgan Kaufmann.

[105] H. Poon and P. Domingos. Sound and efficient inference with probabilistic and deterministic dependencies. In *Proceedings of the Twenty-First National Conference on Artificial Intelligence*, pages 458–463, Boston, MA, 2006. AAAI Press.

[106] H. Poon and P. Domingos. Joint inference in information extraction. In *Proceedings of the Twenty-Second National Conference on Artificial Intelligence*, pages 913–918, Vancouver, Canada, 2007. AAAI Press.

[107] H. Poon and P. Domingos. Joint unsupervised coreference resolution with Markov logic. In *Proceedings of the 2008 Conference on Empirical Methods in Natural Language Processing*, pages 650–659, Honolulu, HI, 2008. ACL.

[108] H. Poon, P. Domingos, and M. Sumner. A general method for reducing the complexity of relational inference and its application to MCMC. In *Proceedings of the Twenty-Third National Conference on Artificial Intelligence*, pages 1075–1080, Chicago, IL, 2008. AAAI Press.

[109] A. Popescul and L. H. Ungar. Structural logistic regression for link analysis. In S. Džeroski, L. De Raedt, and S. Wrobel, editors, *Proceedings of the Second International Workshop on Multi-Relational Data Mining*, pages 92–106, Washington, DC, 2003. ACM Press.

[110] A. Popescul and L. H. Ungar. Cluster-based concept invention for statistical relational learning. In *Proceedings of the Tenth ACM SIGKDD International Conference on Knowledge Discovery and Data Mining*, pages 665–664, Seattle, WA, 2004. ACM Press. DOI: 10.1145/1014052.1014137

[111] A. Puech and S. Muggleton. A comparison of stochastic logic programs and Bayesian logic programs. In *Proceedings of the IJCAI-2003 Workshop on Learning Statistical Models from Relational Data*, pages 121–129, Acapulco, Mexico, 2003. IJCAII.

[112] M. R. Quillian. Semantic memory. In M. L. Minsky, editor, *Semantic Information Processing*, pages 216–270. MIT Press, Cambridge, MA, 1968.

[113] J. R. Quinlan. Learning logical definitions from relations. *Machine Learning*, 5:239–266, 1990. DOI: 10.1007/BF00117105

[114] B. L. Richards and R. J. Mooney. Learning relations by pathfinding. In *Proceedings of the Tenth National Conference on Artificial Intelligence*, pages 50–55, San Jose, CA, 1992. AAAI Press.

[115] B. L. Richards and R. J. Mooney. Automated refinement of first-order Horn-clause domain theories. *Machine Learning*, 19(2):95–131, 1995. DOI: 10.1007/BF01007461

[116] M. Richardson and P. Domingos. Building large knowledge bases by mass collaboration. In *Proceedings of the Second International Conference on Knowledge Capture*, pages 129–137, Sanibel Island, FL, 2003. ACM Press. DOI: 10.1145/945645.945665

[117] M. Richardson and P. Domingos. Markov logic networks. *Machine Learning*, 62:107–136, 2006. DOI: 10.1007/s10994-006-5833-1

[118] S. Riedel. Improving the accuracy and efficiency of MAP inference for Markov logic. In *Proceedings of the Twenty-Fourth Conference on Uncertainty in Artificial Intelligence*, pages 468–475, Helsinki, Finland, 2008. AUAI Press.

[119] S. Riedel and E. Klein. Genic interaction extraction with semantic and syntactic chains. In *Proceedings of the Fourth Workshop on Learning Language in Logic*, pages 69–74, Bonn, Germany, 2005. IMLS.

[120] S. Riedel and I. Meza-Ruiz. Collective semantic role labelling with Markov logic. In *Proceedings of the 2008 Conference on Empirical Methods in Natural Language Processing*, pages 188–192, Honolulu, HI, 2008. ACL.

[121] S. Riezler. *Probabilistic Constraint Logic Programming*. PhD thesis, University of Tubingen, Tubingen, Germany, 1998.

[122] J. A. Robinson. A machine-oriented logic based on the resolution principle. *Journal of the ACM*, 12:23–41, 1965. DOI: 10.1145/321250.321253

[123] D. Roth. On the hardness of approximate reasoning. *Artificial Intelligence*, 82:273–302, 1996. DOI: 10.1016/0004-3702(94)00092-1

[124] D. Roy, C. Kemp, V. K. Mansinghka, and J. B. Tenenbaum. Learning annotated hierarchies from relational data. In *Advances in Neural Information Processing Systems 19*, pages 1210–1217, Cambridge, MA, 2006. MIT Press.

[125] D. M. Roy and L. P. Kaelbling. Efficient Bayesian task-level transfer learning. In *Proceedings of the Twentieth International Joint Conference on Artificial Intelligence*, pages 2599–2604, Hyderabad, India, 2007. AAAI Press.

[126] R. J. Rummel. Dimensionality of nations project: attributes of nations and behavior of nation dyads, 1950 -1965. ICPSR data file. 1999. DOI: 10.3886/ICPSR05409

[127] S. Sanner. *First-Order Decision-Theoretic Planning in Structured Relational Environments*. PhD thesis, University of Toronto, 2008.

[128] V. Santos Costa, D. Page, M. Qazi, , and J. Cussens. CLP(BN): Constraint logic programming for probabilistic knowledge. In *Proceedings of the Nineteenth Conference on Uncertainty in Artificial Intelligence*, pages 517–524, Acapulco, Mexico, 2003. Morgan Kaufmann. DOI: 10.1007/978-3-540-78652-8

[129] T. Sato and Y. Kameya. PRISM: A symbolic-statistical modeling language. In *Proceedings of the Fifteenth International Joint Conference on Artificial Intelligence*, pages 1330–1335, Nagoya, Japan, 1997. Morgan Kaufmann.

[130] S. Schoenmackers, O. Etzioni, and D. S. Weld. Scaling textual inference to the Web. In *Proceedings of the 2008 Conference on Empirical Methods in Natural Language Processing*, pages 79–88, Honolulu, HI, 2008. ACL.

[131] F. Sha and F. Pereira. Shallow parsing with conditional random fields. In *Proceedings of the 2003 Human Language Technology Conference and North American Chapter of the Association for Computational Linguistics*, pages 134–141. Association for Computational Linguistics, 2003. DOI: 10.3115/1073445.1073473

[132] J. Shewchuck. An introduction to the conjugate gradient method without the agonizing pain. Technical Report CMU-CS-94-125, School of Computer Science, Carnegie Mellon University, 1994.

[133] P. Singla and P. Domingos. Discriminative training of Markov logic networks. In *Proceedings of the Twentieth National Conference on Artificial Intelligence*, pages 868–873, Pittsburgh, PA, 2005. AAAI Press.

[134] P. Singla and P. Domingos. Entity resolution with Markov logic. In *Proceedings of the Sixth IEEE International Conference on Data Mining*, pages 572–582, Hong Kong, 2006. IEEE Computer Society Press. DOI: 10.1109/ICDM.2006.65

[135] P. Singla and P. Domingos. Memory-efficient inference in relational domains. In *Proceedings of the Twenty-First National Conference on Artificial Intelligence*, pages 488–493, Boston, MA, 2006. AAAI Press.

[136] P. Singla and P. Domingos. Markov logic in infinite domains. In *Proceedings of the Twenty-Third Conference on Uncertainty in Artificial Intelligence*, pages 368–375, Vancouver, Canada, 2007. AUAI Press.

[137] P. Singla and P. Domingos. Markov logic in infinite domains. Technical report, Dept. Comp. Sci. & Eng., Univ. Washington, 2007.
http://alchemy.cs.washington.edu/papers/singla07/tr.pdf.

[138] P. Singla and P. Domingos. Lifted first-order belief propagation. In *Proceedings of the Twenty-Third National Conference on Artificial Intelligence*, pages 1094–1099, Chicago, IL, 2008. AAAI Press.

[139] P. Singla, H. Kautz, J. Luo, and A. Gallagher. Discovery of social relationships in consumer photo collections using Markov logic. In *2008 CVPR Workshop on Semantic Learning and Applications in Multimedia*, Anchorage, Alaska, 2008. DOI: 10.1109/CVPRW.2008.4563047

[140] A. Srinivasan. The Aleph manual. Technical report, Computing Laboratory, Oxford University, 2000.

[141] B. Taskar, P. Abbeel, and D. Koller. Discriminative probabilistic models for relational data. In *Proceedings of the Eighteenth Conference on Uncertainty in Artificial Intelligence*, pages 485–492, Edmonton, Canada, 2002. Morgan Kaufmann.

[142] M. E. Taylor, A. Fern, and K. Driessens, editors. *Proceedings of the AAAI-2008 Workshop on Transfer Learning for Complex Tasks*. AAAI Press, Chicago, IL, 2008.

[143] M. E. Taylor and P. Stone. Cross-domain transfer for reinforcement learning. In *Proceedings of the Twenty-Fourth International Conference on Machine Learning*, pages 879–886, Corvallis, OR, 2007. ACM Press. DOI: 10.1145/1273496.1273607

[144] L. Torrey, T. Walker, J. Shavlik, and R. Maclin. Relational macros for transfer in reinforcement learning. In *Proceedings of the Seventeenth International Conference on Inductive Logic Programming*, pages 254–268, Corvallis, OR, 2007. Springer. DOI: 10.1007/978-3-540-78469-2

[145] G. G. Towell and J. W. Shavlik. Knowledge-based artificial neural networks. *Artificial Intelligence*, 70:119–165, 1994. DOI: 10.1016/0004-3702(94)90105-8

[146] M. van Otterlo. A survey of reinforcement learning in relational domains. Technical report, University of Twente, 2005.

[147] M. Vilain, J. Burger, J. Aberdeen, D. Connolly, and L. Hirschman. A model-theoretic coreference scoring scheme. In *Proceedings of the 6th conference on message understanding*, pages 45–52, Columbia, MD, 1995. ACL. DOI: 10.3115/1072399.1072405

[148] J. Wang and P. Domingos. Hybrid Markov logic networks. In *Proceedings of the Twenty-Third National Conference on Artificial Intelligence*, pages 1106–1111, Chicago, IL, 2008. AAAI Press.

[149] S. Wasserman and K. Faust. *Social Network Analysis: Methods and Applications*. Cambridge University Press, Cambridge, UK, 1994.

[150] W. Wei, J. Erenrich, and B. Selman. Towards efficient sampling: Exploiting random walk strategies. In *Proceedings of the Nineteenth National Conference on Artificial Intelligence*, pages 670–676, San Jose, CA, 2004. AAAI Press.

[151] M. Wellman, J. S. Breese, and R. P. Goldman. From knowledge bases to decision models. *Knowledge Engineering Review*, 7:35–53, 1992.

[152] B. Wellner, A. McCallum, F. Peng, and M. Hay. An integrated, conditional model of information extraction and coreference with application to citation matching. In *Proceedings of the Twentieth Conference on Uncertainty in Artificial Intelligence*, pages 593–601, Banff, Canada, 2004. AUAI Press.

[153] A. P. Wolfe and D. Jensen. Playing multiple roles: discovering overlapping roles in social networks. In *Proceedings of the ICML-2004 Workshop on Statistical Relational Learning and its Connections to Other Fields*, pages 49–54, Banff, Canada, 2004. IMLS.

[154] F. Wu and D. Weld. Automatically refining the Wikipedia infobox ontology. In *Proceedings of the 17th International World Wide Web Conference*, pages 635–644, Beijing, China, 2008. ACM Press. DOI: 10.1145/1367497.1367583

[155] Z. Xu, V. Tresp, K. Yu, S. Yu, and H.-P. Kriegel. Dirichlet enhanced relational learning. In *Proceedings of the Twenty-Second International Conference on Machine Learning*, pages 1004–1011, Bonn, Germany, 2005. ACM Press. DOI: 10.1145/1102351.1102478

[156] J. S. Yedidia, W. T. Freeman, and Y. Weiss. Generalized belief propagation. In T. Leen, T. Dietterich, and V. Tresp, editors, *Advances in Neural Information Processing Systems 13*, pages 689–695. MIT Press, Cambridge, MA, 2001.

Biography

PEDRO DOMINGOS

Pedro Domingos is Associate Professor of Computer Science and Engineering at the University of Washington. His research interests are in artificial intelligence, machine learning and data mining. He received a PhD in Information and Computer Science from the University of California at Irvine, and is the author or co-author of over 150 technical publications. He is a member of the editorial board of the Machine Learning journal, co-founder of the International Machine Learning Society, and past associate editor of JAIR. He was program co-chair of KDD-2003 and SRL-2009, and has served on numerous program committees. He has received several awards, including a Sloan Fellowship, an NSF CAREER Award, a Fulbright Scholarship, an IBM Faculty Award, and best paper awards at KDD-98, KDD-99 and PKDD-2005.

DANIEL LOWD

Daniel Lowd is a PhD candidate in the Department of Computer Science and Engineering at the University of Washington. His research covers a range of topics in statistical machine learning, including statistical relational representations, unifying learning and inference, and adversarial machine learning scenarios (e.g., spam filtering). He has received graduate research fellowships from the National Science Foundation and Microsoft Research.